Logarithmic Forms and Diophantine Geometry

There is now much interplay between studies on logarithmic forms and deep aspects of arithmetic algebraic geometry. New light has been shed, for instance, on the famous conjectures of Tate and Shafarevich relating to abelian varieties and the associated celebrated discoveries of Faltings establishing the Mordell conjecture. This book gives an account of the theory of linear forms in the logarithms of algebraic numbers with special emphasis on the important developments of the past twenty-five years. The first part concentrates on basic material in transcendental number theory but with a modern perspective including discussion of the Mahler–Manin conjecture, of the Riemann hypothesis over finite fields, of significant new studies on the effective solution of Diophantine problems and of the *abc*-conjecture. The remainder assumes some background in Lie algebras and group varieties and it covers, in certain instances for the first time in book form, more advanced topics including the work of Masser and Wüstholz on zero estimates on group varieties (derived by a new, more algebraic approach that involves Hilbert functions and Poincaré series), the analytic subgroup theorem and its principal applications; these areas reflect substantial original research. The final chapter summarises other aspects of Diophantine geometry including hypergeometric theory and the André–Oort conjecture. A comprehensive bibliography rounds off this definitive survey of effective methods in Diophantine geometry.

ALAN BAKER, FRS, is Emeritus Professor of Pure Mathematics in the University of Cambridge and Fellow of Trinity College, Cambridge. He has received numerous international awards, including, in 1970, a Fields medal for his work in number theory. This is his third authored book: he has edited three others for publication.

GISBERT WÜSTHOLZ is Professor of Mathematics at ETH Zürich. This is his second authored book and he has been involved in the production of three others.

NEW MATHEMATICAL MONOGRAPHS

All the titles listed below can be obtained from good booksellers or from Cambridge University Press. For a complete series listing visit http://www.cambridge.org/uk/series/sSeries.asp?code=NMM

1 M. Cabanes and M. Enguehard *Representation Theory of Finite Reductive Groups*
2 J. B. Garnett and D. E. Marshall *Harmonic Measure*
3 P. M. Cohn *Free Ideal Rings and Localization in General Rings*
4 E. Bombieri and W. Gubler *Heights in Diophantine Geometry*
5 Y. J. Ionin and M. S. Shrikhande *Combinatorics of Symmetric Designs*
6 S. Berhanu, P. D. Cordaro and J. Hounie *An Introduction to Involutive Structures*
7 A. Shlapentokh *Hilbert's Tenth Problem*
8 G. O. Michler *Theory of Finite Simple Groups*
9 A. Baker and G. Wüstholz *Logarithmic Forms and Diophantine Geometry*
10 P. Kronheimer and T. Mrowka *Monopoles and Three-Manifolds*
11 B. Bekka, P. de la Harpe and A. Valette *Kazhdan's Property (T)*

LOGARITHMIC FORMS AND DIOPHANTINE GEOMETRY

A. BAKER
University of Cambridge

G. WÜSTHOLZ
ETH Zentrum, Zürich

CAMBRIDGE
UNIVERSITY PRESS

CAMBRIDGE UNIVERSITY PRESS
Cambridge, New York, Melbourne, Madrid, Cape Town, Singapore, São Paulo

Cambridge University Press
The Edinburgh Building, Cambridge CB2 8RU, UK

Published in the United States of America by Cambridge University Press, New York

www.cambridge.org
Information on this title: www.cambridge.org/9780521882682

First published 2007

Printed in the United Kingdom at the University Press, Cambridge

A catalogue record for this publication is available from the British Library

ISBN 978-0-521-88268-2 hardback

Contents

Preface

This book has arisen from lectures given by the first author at ETH Zürich in the Wintersemester 1988–1989 under the Nachdiplomvorlesung program and subsequent lectures by both authors in various localities, in particular at an instructional conference organised by the DMV in Blaubeuren. Our object has been to give an account of the theory of linear forms in the logarithms of algebraic numbers with special emphasis on the important developments of the past twenty-five years concerning multiplicity estimates on group varieties.

As will be clear from the text there is now much interplay between studies on logarithmic forms and deep aspects of arithmetic algebraic geometry. New light has been shed for instance on the famous conjectures of Tate and Shafarevich relating to abelian varieties and the associated celebrated discoveries of Faltings establishing the Mordell conjecture. We give a connected exposition reflecting these major advances including the first version in book form of the basic works of Masser and Wüstholz on zero estimates on group varieties, the analytic subgroup theorem and their applications. Our discussion here is more algebraic in character than the original and involves, in particular, Hilbert functions in degree theory and Poincaré series as well as the general background of Lie algebras and group varieties. On the other hand, the first three chapters have been written on a more basic level in the style of Baker [25]; since its publication in 1975, the latter has been the classical introduction to transcendence theory, and especially to the subject of logarithmic forms, and it may still be regarded as the standard work in this field. The text here gives in essence a new rendering and updating of Chapters 1 to 5 of [25].

We are most grateful to Camilla Grob for her unstinting help in taking down our lecture notes with a view to publication and to S. Gerig, F. Yan and O. Fasching for their generous assistance in connection with the detailed preparation of the text, in particular with the LaTeX typesetting. We are much indebted to Professor D. W. Masser for reading through a draft of the book prior to publication and for making many detailed and helpful suggestions. Further we thank Professor P. Cohen for reviewing aspects of the book, in particular in connection with Chapter 8. Finally we acknowledge with gratitude the generous support of the Forschungsinstitut at ETH in arranging a variety of visits so that we could complete our work.

A. Baker and G. Wüstholz (Cambridge and Zürich)

1

Transcendence origins

1.1 Liouville's theorem

In 1844 Liouville showed for the first time the existence of transcendental numbers, that is numbers which are not algebraic and so are not roots of any polynomial with integer coefficients [147]. The following approximation theorem by Liouville allowed a certain type of number to be established as transcendental.

Theorem 1.1 (Liouville) *If α is an algebraic number with degree $n > 1$ then, for all rationals p/q (p, $q \in \mathbb{Z}$, $q > 0$), we have*

$$\left|\alpha - \frac{p}{q}\right| > \frac{c}{q^n}$$

for some constant $c = c(\alpha) > 0$ (that is, c is only dependent on α).

Proof. Let $P(x)$ be the minimal polynomial for α (that is the irreducible polynomial P with $P(\alpha) = 0$, with the coefficients of P integers, with the leading coefficient positive and with the greatest common divisor of the coefficients equal to 1). We can assume that α is real and that $|\alpha - p/q| < 1$, for otherwise the theorem is trivially valid. By the mean value theorem we have $P(\alpha) - P(p/q) = (\alpha - p/q)P'(\xi)$ for some ξ between α and p/q. Then ξ belongs to $(\alpha - 1, \alpha + 1)$ and therefore $\left|P'(\xi)\right| < 1/c$ for some $c = c(\alpha) > 0$. Since $P(\alpha) = 0$ we get

$$\left|\alpha - \frac{p}{q}\right| > c\left|P\left(\frac{p}{q}\right)\right|.$$

Since P is irreducible of degree n, $P(p/q) \neq 0$ and $|q^n P(p/q)|$ is an integer, whence $|P(p/q)| \geq 1/q^n$ and the theorem follows. □

Now let us look at some numbers for which this theorem provides a proof of their transcendence.

Example 1.2 *The number*

$$\xi = \sum_{n=1}^{\infty} 10^{-n!}$$

is transcendental.

For let $p_k = 10^{k!} \sum_{n=1}^{k} 10^{-n!}$ and $q_k = 10^{k!}$ for $k = 1, 2, \ldots$; then p_k, q_k are relatively prime rational integers and

$$\left|\xi - \frac{p_k}{q_k}\right| = \sum_{n=k+1}^{\infty} 10^{-n!} < 10^{-(k+1)!} \sum_{n=0}^{\infty} 10^{-n}$$
$$= \frac{10}{9} q_k^{-(k+1)} < q_k^{-k}.$$

Since k tends to infinity there cannot exist a constant c, as in the theorem, only depending on ξ. Therefore ξ is transcendental. Further, as immediate consequences of Liouville's theorem, we have the following.

Example 1.3 *Any non-terminating decimal of the type*

$$0.a_1 0 \cdots 0 a_2 0 \cdots 0 a_3 0 \cdots,$$

in which blocks of zeros increase in length sufficiently rapidly, is transcendental. Similarly any continued fraction in which the partial quotients increase sufficiently rapidly is transcendental.

In 1906 Maillet published the first book on transcendental numbers [159]. He showed here, amongst other things, that there exist transcendental numbers whose continued fractions have bounded partial quotients.

Example 1.4 *Continued fractions of the type*

$$[1, \ldots, 1, a_1, 1, \ldots, 1, a_2, 1, \ldots, 1, a_3, 1, \ldots]$$

are transcendental, where $a_i \neq 1$ and the number of repeated partial quotients increases sufficiently rapidly.

The subject of continued fractions of Maillet type was taken up by Baker [11] and it continues to be of research interest (see e.g. [180]). Maillet's proof was based on an approximation theorem with quadratic irrationals. In 1961, Güting [122] obtained an elegant theorem of this kind relating to numbers of arbitrary degree. In order to state the result we need the concept of the height of an algebraic number; in fact some notion of height occurs throughout our text. Let α be an algebraic number and let the minimal polynomial for α be

$$P(x) = a_0 x^n + a_1 x^{n-1} + \cdots + a_n .$$

Definition 1.5 *The (classical) height of α is given by*

$$H(\alpha) = \max(|a_0|, \ldots, |a_n|) .$$

Now let α and β be distinct algebraic numbers with heights a and b, and let l, m be the degrees of β over $\mathbb{Q}(\alpha)$ and α over $\mathbb{Q}(\beta)$ respectively. Then Güting's theorem reads as follows.

Theorem 1.6 *We have*

$$|\alpha - \beta| \gg a^{-l} b^{-m} .$$

Here we are using Vinogradov's notation: by $f \gg g$ for functions f, g we mean $f > cg$ for some positive constant c and similarly by $f \ll g$ we mean $f < cg$. The constant c in Theorem 1.6 is effective and for an explicit expression in terms of l and m see [122].

Proof of Theorem 1.6. Güting's argument is essentially a straightforward generalisation of Liouville's. It depends on the fact that

$$|a_0^l b_0^m N(\alpha - \beta)| \geq 1 ,$$

where a_0 and b_0 are the leading coefficients in the minimal polynomials for α and β, and N denotes the field norm with respect to $\mathbb{Q}(\alpha, \beta)$. The field conjugates $\alpha_j - \beta_j$ of $\alpha - \beta$ have absolute value at most $(1 + |\alpha_j|)(1 + |\beta_j|)$ and estimates for $a_0^l \prod(1 + |\alpha_j|)$ and $b_0^m \prod(1 + |\beta_j|)$

in terms of the heights a and b, where the products are taken over all field conjugates, date back to Landau; see [33, §2]. In fact Theorem 4.2 of LeVeque's book [145] shows that the expressions are at most $6^n a^l$ and $6^n b^m$ respectively where n denotes the degree of $\mathbb{Q}(\alpha, \beta)$ and Theorem 1.6 follows. □

A much deeper result in the context of Liouville's theorem was discovered by Thue [243] in 1909 and Thue's work was subsequently developed in important papers by Siegel [226], Schneider [212], Dyson [81], Gelfond [108] and Roth [204]. Let α be an algebraic number with degree $n > 1$ and consider the inequality

$$\left|\alpha - \frac{p}{q}\right| > \frac{c}{q^{\varkappa}}$$

for $c = c(\alpha, \varkappa) > 0$ and p, q rational integers. Then Thue showed that $c(\alpha, \varkappa)$ exists for $\varkappa > \frac{1}{2}n + 1$. The result was sharpened by Siegel to $\varkappa > s + n/(s+1)$ for any positive integer s, in particular to $\varkappa > 2\sqrt{n}$, and this was further improved by Dyson and Gelfond independently to $\varkappa > \sqrt{2n}$. Finally Roth showed that there exists $c(\alpha, \varkappa) > 0$ for any $\varkappa > 2$ and, by continued fraction theory for example, this is best possible.

Theorem 1.7 (Thue–Siegel–Roth) *If* $\varkappa > 2$ *then there exists* $c(\alpha, \varkappa) > 0$ *such that the above inequality holds for all rationals* p/q $(q > 0)$.

Thue was motivated by studies on Diophantine equations and one of the main applications of his result was a demonstration of the finiteness of the number of solutions of the equation $F(x, y) = m$ where F is an irreducible binary form with integer coefficients and degree at least 3 (see Section 3.3). Siegel's sharpening led to his famous theorem that there are only finitely many integer points on any algebraic curve of genus at least 1. The works of Thue and Siegel were based on the construction of a polynomial in two variables by means of the box principle and they yielded an estimate for the number of solutions to the equations in question. But they did not furnish an estimate for the sizes of the solutions and so they did not enable one to actually solve the equations. The reason lay in the ineffectiveness of the constant c in Theorem 1.7 and its earlier versions subsequent to that of Liouville; it arises from a purely hypothetical assumption at the beginning of the proof that α has at least one good approximation p/q with large q. The first effective improvement on Liouville's theorem for some particular algebraic numbers was obtained by Baker [12] using a method involving hypergeometric functions. As an example he showed [13] that

$$\left|\sqrt[3]{2} - \frac{p}{q}\right| > 10^{-6}\frac{1}{q^{2.955}}.$$

This immediately yields a bound in terms of m for all integer solutions of the Diophantine equation $x^3 - 2y^3 = m$ and indeed it enables one to solve the equation completely for any reasonably sized m. Many other examples of this type relating to approximation to fractional powers of rationals can be given; see especially [69]. However, it was not until Baker's development of the theory of linear forms in the logarithms of algebraic numbers [15] that one was able to give the first general effective improvement on Liouville's theorem. The latter theory and its ramifications will be the main theme of this book.

Before closing this section it should be mentioned that Bombieri [47] (see also the discussion in [48]) has recently succeeded in obtaining an alternative approach to questions on effective improvements on Liouville's theorem. His work is based on the original Thue–Siegel technique and surprisingly he shows that this can be made effective. But the method based on the theory of logarithmic forms would seem at present to be stronger.

1.2 The Hermite–Lindemann theorem

In 1873 Hermite [127] proved that e is transcendental. His proof was based on Padé approximants to $e^x, \ldots, e^{nx}$. Lindemann [146] extended Hermite's method to $e^{\alpha_1 x}, \ldots, e^{\alpha_n x}$ and showed thereby in 1882 that π is transcendental (see Section 6.3 for further historical details). In fact Lindemann proved a much more general result which includes the transcendence of e and π as special cases.

Theorem 1.8 *Whenever $\alpha_0, \ldots, \alpha_n$ are distinct algebraic numbers and $\beta_0, \ldots, \beta_n$ are non-zero algebraic numbers we have*

$$\beta_0 e^{\alpha_0} + \cdots + \beta_n e^{\alpha_n} \neq 0.$$

Plainly the transcendence of e follows on taking $\alpha_j = j$ and the β as integers or simply on taking $n = 1$, $\alpha_0 = 0$, $\alpha_1 = 1$ and $\beta_1 = -1$. Further, the transcendence of π follows from Euler's equation $e^{i\pi} = -1$. It is also readily seen that Theorem 1.8 implies the transcendence of e^{α} and $\log \alpha$ for algebraic $\alpha \neq 0, 1$, and also the transcendence of the trigonometric functions $\cos \alpha$, $\sin \alpha$ and $\tan \alpha$ for algebraic $\alpha \neq 0$.

Proof of Theorem 1.8. A proof of the theorem is given in [25, Ch. 1, §3]. We shall not repeat the details here but shall give instead a demonstration of the transcendence of π following the same method.

Accordingly suppose that π is algebraic. On defining $\vartheta = i\pi$ and using Euler's identity $e^{i\pi} = -1$ we get $e^{\vartheta} = -1$ whence

$$(e^{\vartheta_1} + 1) \cdots (e^{\vartheta_d} + 1) = 0,$$

where $\vartheta_1, \ldots, \vartheta_d$ denote the conjugates of ϑ. On expanding the left-hand side we obtain a sum of 2^d terms e^{Θ}, where

$$\Theta = \varepsilon_1 \vartheta_1 + \cdots + \varepsilon_d \vartheta_d$$

and $\varepsilon_j = 0$ or 1; we suppose that precisely n of the numbers Θ are non-zero and we denote these by $\alpha_1, \ldots, \alpha_n$. We have then

$$b_0 + b_1 e^{\alpha_1} + \cdots + b_n e^{\alpha_n} = 0,$$

where b_0 is the positive integer $2^d - n$, where $b_1 = \cdots = b_n = 1$ and $\alpha_1, \ldots, \alpha_n$ are algebraic numbers such that $\mathbb{Q}(\alpha_1, \ldots, \alpha_n)$ is a Galois field, that is $\alpha_1, \ldots, \alpha_n$ can be written as complete sets of conjugates. We proceed to show that the equation is impossible; indeed we shall prove this under the more general assumption that the b are arbitrary integers such that for each complete set of conjugates $\alpha_{k_1}, \ldots, \alpha_{k_m}$ the corresponding $b_{k_1}, \ldots, b_{k_m}$ are equal. The latter assumption and the Galois condition hold trivially on taking $\alpha_j = j$ and so our result will then include the transcendence of e.

We define

$$I(t) = \int_0^t e^{t-u} f(u)\, du,$$

where $f(x) = l^{np} x^{p-1} (x - \alpha_1)^p \cdots (x - \alpha_n)^p$; here p denotes a large prime and l is any positive integer such that $l\alpha_1, \ldots, l\alpha_n$ are algebraic integers. Now by iteration of partial integration we get

$$I(t) = e^t \sum_{j=0}^{m} f^{(j)}(0) - \sum_{j=0}^{m} f^{(j)}(t),$$

where $m = (n+1)p - 1$ and $f^{(j)}$ is the jth derivative of f. Let $\bar{f}$ be the polynomial obtained from f by replacing each coefficient of f with its absolute value; then

$$|I(t)| \leq |t| e^{|t|} \bar{f}(|t|).$$

We shall compare estimates for

$$J = b_1 I(\alpha_1) + \cdots + b_n I(\alpha_n).$$

By the exponential equation and the expression for $I(t)$ above we have

$$\begin{aligned} J &= \sum_{k=1}^{n} b_k e^{\alpha_k} \sum_{j=0}^{m} f^{(j)}(0) - \sum_{k=1}^{n} b_k \sum_{j=0}^{m} f^{(j)}(\alpha_k) \\ &= -b_0 \sum_{j=0}^{m} f^{(j)}(0) - \sum_{j=0}^{m} \sum_{k=1}^{n} b_k f^{(j)}(\alpha_k). \end{aligned}$$

Now we know by our Galois assumption that J remains fixed under the automorphisms of $\overline{\mathbb{Q}}$ (algebraic closure of $\mathbb{Q}$) and is therefore a rational integer (note that the coefficients of f are symmetric in the α_j). By the definition of f we have $f^{(j)}(\alpha_k) = 0$ for $j < p$ and $f^{(j)}(0) = 0$ for $j < p - 1$ and

$$f^{(p-1)}(0) = (-1)^n \, (p-1)! \, (l^n \alpha_1 \cdots \alpha_n)^p.$$

If p is sufficiently large (not dividing $l^n\alpha_1\cdots\alpha_n$) then $f^{(p-1)}(0)$ is divisible by $(p-1)!$ but not by $p!$. Hence we obtain

$$J = -b_0 f^{(p-1)}(0) - \left(b_0 \sum_{j=p}^{m} f^{(j)}(0) + \sum_{j=p}^{m}\sum_{k=1}^{n} b_k f^{(j)}(\alpha_k)\right).$$

The expression in brackets is divisible by $p!$. Thus J is divisible by $(p-1)!$ and not by $p!$ whence $|J| \geq (p-1)!$.

On the other hand we have $\bar{f}(|\alpha_j|) < C^p$ for some constant C independent of p and by the estimate for $I(t)$ we get $|J| < c^p$ for some constant c independent of p. This gives a contradiction, as required. □

A classical form of Lindemann's theorem which is equivalent to Theorem 1.8 is the following.

Theorem 1.9 (Lindemann) *If $\alpha_1, \ldots, \alpha_n$ are algebraic numbers linearly independent over $\mathbb{Q}$ then*

$$e^{\alpha_1}, \ldots, e^{\alpha_n}$$

are algebraically independent.

Proof. For any polynomial $P(x_1, \ldots, x_n)$ with algebraic coefficients not all zero the equation $P(e^{\alpha_1}, \ldots, e^{\alpha_n}) = 0$ implies that

$$\beta_1 e^{\alpha'_1} + \cdots + \beta_m e^{\alpha'_m} = 0$$

with algebraic $\alpha'_1, \ldots, \alpha'_m$, which are different linear combinations of $\alpha_1, \ldots, \alpha_n$ with integer coefficients and with $\beta_1, \ldots, \beta_m$ not all zero. This contradicts Theorem 1.8. Conversely Theorem 1.8 is an immediate consequence of Theorem 1.9 on observing that if $\alpha_1, \ldots, \alpha_n$ are algebraic and distinct and if $\omega_1, \ldots, \omega_d$ is an integral basis for $\mathbb{Q}(\alpha_1, \ldots, \alpha_n)$ then, for some positive integer l, each α_j is expressible as a linear combination of $\omega_1/l, \ldots, \omega_d/l$ with integer coefficients. □

Studies arising from the Hermite–Lindemann theory have yielded measures of irrationality and transcendence for e and π and other related numbers. Mahler [155] obtained an especially striking result in this context, namely

$$|\pi - p/q| > 1/q^{42}$$

valid for all rationals p/q ($q > 1$); for further developments see D. V. and G. V. Chudnovsky [66]. In connection with e we have the continued fraction

$$e = [2, 1, 2, 1, 1, 4, 1, 1, 6, \ldots]$$

and this implies a very sharp measure of irrationality (see Davis [77]); in particular we have

$$|e - p/q| > c/q^{2+\varepsilon}$$

for any $\varepsilon > 0$, where $c = c(\varepsilon)$ is an effective positive constant, and here one can replace ε by a function $\varepsilon(q) \to 0$ as $q \to \infty$. More generally we have the measure of transcendence

$$|e - \alpha| > c/H^{n+1+\varepsilon}$$

valid for all algebraic numbers α with degree n and height H, where c is a positive constant depending on n and ε. In fact much more is known about the approximation properties concerning e and its powers. Indeed it has been proved (see [14; 25, Ch. 10]) that there are only finitely many non-zero integers $a_0, a_1, \ldots, a_n$ such that

$$|a_1 \cdots a_n|^{1+\varepsilon} \left|a_0 + a_1 e + \cdots + a_n e^n\right| < 1$$

or, equivalently, that there are only finitely many positive integers q and integers $p_1, \ldots, p_n$ such that

$$q^{1+\varepsilon} |qe - p_1| \cdots \left|qe^n - p_n\right| < 1.$$

In other words, the vector $(e, e^2, \ldots, e^n)$ is not very well multiplicatively approximated in the language of Kleinbock and Margulis [130]. The same holds for any vector $(e^{r_1}, \ldots, e^{r_n})$ with distinct non-zero rationals $r_1, \ldots, r_n$.

1.3 The Siegel–Shidlovsky theory

In a fundamental paper of 1929, Siegel [228] introduced the concept of an E-function and established the algebraic independence at algebraic arguments of the values of E-functions subject to them satisfying linear differential equations of the first or second order. Shidlovsky [221] succeeded in 1954 in generalising Siegel's method to differential equations of arbitrary order and many valuable results have followed. For a discussion of the Siegel–Shidlovsky theory see [25, Ch. 11]; we shall give here only a brief indication of some illustrative details.

Siegel defined an E-function as a series

$$\sum_{n=0}^{\infty} a_n \frac{x^n}{n!},$$

where $a_0, a_1, \ldots, a_i, \ldots$ are elements of an algebraic number field $\mathbb{K}$ such that, for some sequence of positive rational integers $b_0, b_1, \ldots$ and for any $\varepsilon > 0$, the numbers $b_n a_0, b_n a_1, \ldots, b_n a_n$ and b_n are elements of the ring $\mathcal{O}_{\mathbb{K}}$ of algebraic integers of $\mathbb{K}$ with sizes $\ll n^{\varepsilon n}$ where the implied constant depends only on ε; here by the size of an algebraic integer we mean the maximum of the absolute values of its conjugates. Siegel noted that sums, products, derivatives and integrals of E-functions are again E-functions. He considered E-functions $E_i(x)$ $(1 \leq i \leq n)$ that satisfy a system of linear differential equations

$$y_i' = \sum_{j=1}^{n} f_{ij}(x) y_j \quad (1 \leq i \leq n),$$

where the f_{ij} are rational functions over $\mathbb{K}$. In view of the work of Shidlovsky [221] referred to above we have the following theorem.

Theorem 1.10 (Siegel–Shidlovsky) *Let α be an algebraic number in $\mathbb{K}$ distinct from the poles of the f_{ij}. If the functions $E_1(x), \ldots, E_n(x)$ are algebraically independent over $\mathbb{K}(x)$ then $E_1(\alpha), \ldots, E_n(\alpha)$ are algebraically independent over $\mathbb{K}$.*

Apart from [25, Ch. 11] cited above, other references for the Siegel–Shidlovsky theory are Mahler's tract [158] and Shidlovsky's book [222]. Note that Theorem 1.10 includes Theorem 1.9 of Lindemann as a special case. Indeed one has simply to take $E_j(x) = e^{\alpha_j x}$ $(j = 1, \ldots, n)$; these are clearly E-functions satisfying differential equations as above with $f_{ij}(x) = \alpha_j$ when $i = j$ and 0 otherwise. Siegel himself gave the example

$E_1(x) = J_0(x), E_2(x) = J_0'(x)$ where $J_0(x)$ is the Bessel function, that is

$$J_0(x) = \sum_{n=0}^{\infty} \frac{(-1)^n}{(n!)^2}\left(\frac{x}{2}\right)^{2n},$$

and he deduced that $J_0(\alpha)$ and $J_0'(\alpha)$ are algebraically independent for any non-zero algebraic number α. In particular, as he observed, this implies that the continued fraction $[1, 2, 3, \ldots]$ is transcendental. Many further examples can be found in the literature.

In order to illustrate the method of proof of Theorem 1.10 we shall show, in the case $\mathbb{K} = \mathbb{Q}$, that if the functions $E_1(x), \ldots, E_n(x)$ are linearly independent over $\mathbb{K}(x)$ then $E_1(\alpha), \ldots, E_n(\alpha)$ are linearly independent over $\mathbb{K}$. This includes the transcendence of e as one sees by taking $E_j(x) = e^{jx}$ $(1 \leq j \leq n)$ and $\alpha = 1$. The result for general $\mathbb{K}$ with degree d is that the maximum number of elements $E_1(\alpha), \ldots, E_n(\alpha)$ that are linearly independent over $\mathbb{K}$ is at least n/d (see [222, Ch. 3, §11]). Theorem 1.10 follows readily from the latter on applying the result to power products $(E_1(x))^{l_1} \cdots (E_n(x))^{l_n}$ of the given E-functions.

Proof of Theorem 1.10. Suppose $\varepsilon > 0$ and let r be a large rational integer. From [25, Ch. 11, Lemma 4] there exist rational integers q_{ij} $(1 \leq i, j \leq n)$ with $\det(q_{ij}) \neq 0$ (for details of the construction see below) such that

$$\left|\sum_{i=1}^{n} q_{ij} E_i(\alpha)\right| < (r!)^{-n+1+\varepsilon n} \quad (1 \leq j \leq n)$$

and

$$|q_{ij}| < (r!)^{1+\varepsilon}.$$

Now suppose that $E_1(\alpha), \ldots, E_n(\alpha)$ are linearly dependent over $\mathbb{Q}$ so that

$$p_1 E_1(\alpha) + \cdots + p_n E_n(\alpha) = 0,$$

where the coefficients are rational integers, not all zero. Since the determinant of the q_{ij} is not 0, it follows that there exist $n-1$ of the forms

$$L_j = \sum_{i=1}^{n} q_{ij} E_i(\alpha) \quad (1 \le j \le n)$$

which together with the linear form above in the p_i make up a linearly independent set. Without loss of generality we can suppose that they are given by $L_2, \ldots, L_n$. We shall suppose also, as we may, that $E_1(\alpha) \neq 0$. Let D be the determinant of order n given by

$$D = \begin{vmatrix} p_1 & q_{12} & \cdots & q_{1n} \\ p_2 & q_{22} & & q_{2n} \\ \vdots & & \ddots & \vdots \\ p_n & q_{n2} & \cdots & q_{nn} \end{vmatrix}.$$

By construction, D is a non-zero rational integer and so we have $|D| \ge 1$. On the other hand, by taking linear combinations of rows, we see that

$$E_1(\alpha) D = \begin{vmatrix} 0 & L_2 & \cdots & L_n \\ p_2 & q_{22} & & q_{2n} \\ \vdots & & \ddots & \vdots \\ p_n & q_{n2} & \cdots & q_{nn} \end{vmatrix}.$$

The elements in the first row on the right are $\ll (r!)^{-n+1+\varepsilon n}$, the remaining elements are $\ll (r!)^{1+\varepsilon}$ and thus we obtain

$$|D| \ll (r!)^{(1+\varepsilon)(n-2)-n+1+\varepsilon n} \le (r!)^{-1+2\varepsilon n}.$$

Hence if $\varepsilon < 1/(2n)$ we have a contradiction for r sufficiently large. We conclude that $E_1(\alpha), \ldots, E_n(\alpha)$ are linearly independent over $\mathbb{Q}$, as required. □

To obtain the integers q_{ij} indicated in the proof above one first constructs polynomials $P_i(x)$ $(1 \le i \le n)$, not all identically zero, with degrees at most r and with integer coefficients having absolute values at most $(r!)^{1+\varepsilon}$ (smaller ε than above) such that

$$\sum_{i=1}^{n} P_i(x) E_i(x) = \sum_{m=M}^{\infty} \varrho_m x^m,$$

where $|\varrho_m| < (r!)(m!)^{-1+\varepsilon}$ and $M = n(r+1) - 1 - [\varepsilon r]$. This is established by means of a version of a famous lemma of Siegel relating to the existence of a small non-trivial solution to a system of linear Diophantine equations (see Section 1.4); a lemma of this kind occurs in fact throughout transcendence theory. One now differentiates the power series equation above n times so as to give a set of polynomials $P_{ij}(x)$ satisfying an equation of the form

$$\sum_{i=1}^{n} P_{ij}(x) E_i(x) = \sum_{m=M}^{\infty} \varrho_{mj} x^m .$$

The rationals $p_{ij} = P_{ij}(\alpha)$ have a common denominator $a \le c^r$ for some constant c independent of r and on taking $q_{ij} = ap_{ij}$ we obtain integers as in [25, Ch. 11, Lemma 4] except for the condition $\det(q_{ij}) \ne 0$.

To derive the latter one differentiates the power series equation further, following Siegel, so as to give a new set of polynomials $P_{i,J(j)}(x)$ where the $J(j)$ $(1 \le j \le n)$ are distinct suffixes $\ll \varepsilon r$. The q_{ij} are then defined in terms of the new set and the estimates are essentially the same as before. But the suffixes $J(j)$ can be chosen so that $\det(q_{ij}) \ne 0$. The latter depends on the fact that the determinant $\Delta(x) = \det(P_{ij}(x))$, though possessing a zero at $x = 0$ of high order, does not vanish identically; it is a consequence of the assumed linear independence of $E_1(x), \ldots, E_n(x)$ over $\mathbb{Q}(x)$ (or more generally $\mathbb{K}(x)$) together with the recurrence relations satisfied by the $P_{ij}(x)$ by virtue of the differential equations. Siegel obtained the result for differential equations of the first and second orders and it was Shidlovsky's major contribution to the subject to establish the result for equations of arbitrary order. We refer again to the texts cited above for details.

1.4 Siegel's lemma

As mentioned in Section 1.3, there is a basic lemma that occurs throughout transcendence theory on the solution of linear Diophantine equations. It was implicit in the work of Thue [243] of 1909 and it was given explicitly by Siegel [228] in his work on E-functions in 1929.

Lemma 1.11 *Let M, N be integers with $N > M > 0$. For each integer j with $1 \le j \le N$ let a_{ij} $(1 \le i \le M)$ be integers with absolute values at most A_i (≥ 1). Then there exist integers $x_1, \ldots, x_N$, not all zero, with absolute values at most*

$$X = \prod_{i=1}^{M} (NA_i)^{1/(N-M)},$$

such that

$$\sum_{j=1}^{N} a_{ij} x_j = 0 \quad (1 \le i \le M).$$

Proof. Let $B = [X]$. There are $(B+1)^N$ sets $x_1, \ldots, x_N$ of integers with $0 \le x_j \le B$ $(1 \le j \le N)$ and we have

$$-V_i B \le y_i \le W_i B \quad (1 \le i \le M),$$

where $y_i = \sum_{j=1}^{N} a_{ij} x_j$ and $-V_i$, W_i are the sums of the negative and positive a_{ij} $(1 \le j \le N)$ respectively. Thus $W_i + V_i \le NA_i$. Hence there are at most

$$\prod_{i=1}^{M} (NA_i B + 1)$$

sets $y_1, \ldots, y_M$. Since, by the definition of B,

$$(B+1)^{(N-M)} > \prod_{i=1}^{M} (NA_i)$$

and, by assumption, $A_i \ge 1$ we have

$$(B+1)^N > \prod_{i=1}^{M} (NA_i B + 1).$$

Thus there are two sets $x_1, \ldots, x_N$ which give the same set $y_1, \ldots, y_M$ and their difference gives the required solution. □

The version of the lemma given above appears in Cassels' book [63] and it is a little sharper than Siegel's original result; in place of the value of X Siegel gave the bound $(NA)^{M/(N-M)} + 1$, where A is the maximum of the A_i $(1 \leq i \leq M)$. A generalisation of the lemma to algebraic number fields occurs in Schneider's book [216] (see also [25, Ch. 5, §3]) and there is a further version in Dobrowolski's paper [78] but the most comprehensive and precise result to date is due to Bombieri and Vaaler [50]. Let $\mathbb{K}$ be an algebraic number field with degree d and discriminant Δ and let $\mathcal{O}_{\mathbb{K}}$ be the ring of algebraic integers in $\mathbb{K}$.

Lemma 1.12 *Suppose that $N > M$ and let*

$$L_i = \sum_{j=1}^{N} a_{ij} x_j \quad (1 \leq i \leq M)$$

be linear forms with coefficients in $\mathbb{K}$. There exist elements $x_1, \ldots, x_N$ in $\mathcal{O}_{\mathbb{K}}$, not all zero, such that $L_i(x_1, \ldots, x_N) = 0$ $(1 \leq i \leq M)$ and

$$H(\mathbf{x}) \leq |\Delta|^{1/2} \prod_{i=1}^{M} \left(N^{d/2} H(L_i)\right)^{1/(N-M)}.$$

Here $H(\mathbf{x})$ and $H(L_i)$ denote the Weil heights of $\mathbf{x} = (x_1, \ldots, x_n)$ and L_i respectively, that is

$$H(\mathbf{x}) = \prod_{v} \left(\max_{j} |x_j|_v\right), \quad H(L_i) = \prod_{v} \left(\max_{j} |a_{ij}|_v\right),$$

where the products are over all places v of $\mathbb{K}$; we are using the normalised absolute values, that is

$$|p|_v = p^{-[\mathbb{K}_v : \mathbb{Q}_v]} \quad \text{if} \quad v | p, \quad |x|_v = |x|^{[\mathbb{K}_v : \mathbb{Q}_v]} \quad \text{if} \quad v | \infty,$$

where $\mathbb{K}_v$ is the completion of $\mathbb{K}$ at v and $|x|$ is the real or complex absolute value corresponding to $\mathbb{K}_v$.

Though we shall not need to use Lemma 1.12 later in the book, we remark nonetheless that the proof is based on Minkowski's convex body theorem in the Geometry of Numbers rather than the box principle as in Lemma 1.11 and the approach is through the theory of adeles in algebraic number fields. In particular it involves an adelic version of a theorem of Minkowski on the products of successive minima (cf. a theorem of McFeat [175]). The work originates from Vaaler [247] in which he proved a conjecture of Good to the effect that if C_N is a cube of volume 1 centred at the origin in $\mathbb{R}^N$ and if P_M is any M-dimensional linear subspace of $\mathbb{R}^N$ then $C_N \cap P_M$ has M-dimensional volume ≥ 1. The conclusion of Bombieri and Vaaler [50] reduces to

$$|x_j| < \left(D^{-1}|\det(AA^T)|^{1/2}\right)^{1/(N-M)} \qquad (1 \leq j \leq N)$$

under the hypotheses of Lemma 1.11, where $A = (a_{ij})$ is assumed to have rank M and D is the greatest common divisor of the minors of order M in A. This yields in particular a sharpening from N to $N^{1/2}$ in the classical bounds.

1.5 Mahler's method

A general method for establishing the transcendence of values of functions satisfying functional equations was described by Mahler in 1929 (see [151, 152, 153]). The area was neglected for many years and it only came to fruition after Mahler himself recalled his early work in 1969 [156]. The latter paper gave rise to studies on finite automata (Loxton, van der Poorten), topological dynamics (Morse, Hedlund), harmonic analysis, fractals and statistical mechanics (Allouche, Mendès-France) and on quasi-crystals (Bombieri, J. E. Taylor); for references see Loxton and van der Poorten [150], Loxton [149], Nishioka [188] and for some related irrationality problems see Erdős [85]. We illustrate the method by proving one of the basic results of Mahler.

Theorem 1.13 *The Fredholm series $\sum_{n=0}^{\infty} z^{l^n}$ is transcendental for any algebraic number $z = \alpha$ with $0 < |\alpha| < 1$ and any integer $l \geq 2$.*

Proof. Suppose on the contrary that $f(\alpha)$ is algebraic where $f(z)$ denotes the above series. By $c, c_1, c_2, \ldots$ we shall signify positive real numbers which depend only on f and α. Further we shall denote by k a large rational integer, that is, a parameter $> c$ as above, and by $C, C_1, C_2, \ldots$ positive real numbers which depend only on k.

We begin by establishing the existence of rational integers $p(\lambda, \mu)$, not all zero, with absolute values at most C such that the function

$$\varphi(z) = \sum_{\lambda=0}^{L} \sum_{\mu=0}^{L} p(\lambda, \mu)\, z^{\lambda}\, (f(z))^{\mu}$$

satisfies

$$\varphi^{(j)}(0) = 0 \quad (0 \le j \le k),$$

where $L = 2\left[\sqrt{k}\,\right]$ and $\varphi^{(j)}$ denotes the jth derivative of φ. Indeed we have

$$\varphi^{(j)}(0) = j! \sum p(\lambda, \mu)\, q(j, \lambda, \mu),$$

where the sum is over all λ, μ with $0 \le \lambda, \mu \le L$ and with $\lambda + \mu \le j$. Here $q(j, \lambda, \mu)$ is the Taylor coefficient of $z^{j-\lambda}$ in the expansion of $(f(z))^{\mu}$ at $z = 0$ and, since the Taylor coefficients of f are integers, it is an integer. We have to solve k linear equations in the $(L+1)^2$ unknowns $p(\lambda, \mu)$ and this can be done since $(L+1)^2 > k$; note that there is no need here for a Siegel lemma. Further, $\varphi(z)$ does not vanish identically since $f(z)$ is a transcendental function; the latter follows from the fact that $z = e^{2\pi i(s/l^m)}$ is a pole of $f(z)$ for all integers s, m since $z^{l^n} = 1$ whenever $m < n$, whence the function has a dense set of singularities on the circle $|z| = 1$. Thus $\varphi(z) \sim az^r$ as $z \to 0$ for some $a \ne 0$ and some integer $r \ge k$. This gives

$$0 < \left|\varphi\left(\alpha^{l^h}\right)\right| < C_1 |\alpha|^{kl^h} < e^{-c_2 kl^h}$$

for all integers $h \ge C_2$.

On the other hand we have

$$f\left(\alpha^{l^h}\right) = f(\alpha) - \alpha - \alpha^l - \cdots - \alpha^{l^{h-1}},$$

and hence $\varphi(\alpha^{l^h})$ is an algebraic number with degree at most c_3, a number depending only on f and α. Further, since from the functional equation each conjugate of $f(\alpha^{l^h})$ has absolute value $< c_4^{l^h}$, it follows that each conjugate of $\varphi(\alpha^{l^h})$, obtained by allowing $\alpha, f(\alpha)$ to run through their respective conjugates, has absolute value at most $C_3 c_5^{Ll^h}$. The same estimate plainly holds for a denominator of $\varphi(\alpha^{l^h})$, that is a positive integer

a such that $a\varphi(\alpha^{l^h})$ is an algebraic integer, and hence on taking norms we get

$$\left|\varphi\left(\alpha^{l^h}\right)\right| > e^{-c_6 L l^h}.$$

Here we are using

$$\left|\text{Norm}\left(a\varphi\left(\alpha^{l^h}\right)\right)\right| \geq 1$$

and the fact that each conjugate of $a\varphi(\alpha^{l^h})$ is less than $e^{c_7 L l^h}$ if h is large enough. Since $c_6 L = 2c_6\left[\sqrt{k}\right] < c_2 k$ for $k > c$ sufficiently large we have a contradiction as $h \to \infty$ and this proves Theorem 1.13. □

Some exciting new work of Barré-Sirieix, Diaz, Gramain and Philibert, which can be considered as a variant of the method of Mahler, appeared in 1996 [36]. They succeeded in proving a conjecture of Mahler and Manin on the function

$$J(z) = 1/z + 744 + c_1 z + c_2 z^2 + \cdots$$

with $c_1, c_2, \ldots$ integers, defined by the relation $j(z) = J(e^{2\pi i z})$, where $j(z)$ is the classical elliptic modular function.

Theorem 1.14 *$J(\alpha)$ is transcendental for all algebraic numbers α with $0 < |\alpha| < 1$. The result holds analogously in the p-adic setting, that is for all algebraic α with $0 < |\alpha|_p < 1$.*

Essential use is made throughout the proof of the fact that $J(z)$ satisfies an equation $\Phi_n(J(z), J(z^n)) = 0$ where Φ_n is the modular polynomial of order n. An estimate for the degree of Φ_n is classical and estimates for the sizes of the coefficients go back to Mahler [157] (see also Cohen [72]). The proof begins with the construction by means of a Siegel lemma of an auxiliary function F with a zero to a high order, say N, at the origin; it is given by a polynomial in z and $zJ(z)$ with integer coefficients. Assuming that $J(\alpha)$ is algebraic we see from the modular equation that $J(\alpha^n)$ is also algebraic for all positive integers n. Since the function $zJ(z)$ and hence also $F(z)$ is analytic in the unit disc, it follows that there exists a least integer s such that $F(\alpha^s) \neq 0$. On recalling that N is large and applying the maximum-modulus principle one gets a bound for s. Then from the fact that $F(\alpha^s)$ is algebraic one obtains by a Liouville type estimate a lower bound for $|F(\alpha^s)|$. But, since again N is large, a comparison with an upper bound derived from the Taylor expansion of F gives a contradiction.

The p-adic analogue in Theorem 1.14 was conjectured by Manin in 1971 and, as Barré-Sirieix *et al.* mention in their work, there are applications to the theory of elliptic curves and p-adic L-functions; in particular they refer to a case of a conjecture of Mazur, Tate and Teitelbaum. Nesterenko [187, Ch. 3] derived other important developments from the proof of Theorem 1.14. They relate to modular, elliptic and theta-functions. To give the most basic instance, let $P(z)$, $Q(z)$, $R(z)$ be the Ramanujan functions defined by the relations

$$E_2(\tau) = P(e^{2\pi i\tau}), \quad E_4(\tau) = Q(e^{2\pi i\tau}), \quad E_6(\tau) = R(e^{2\pi i\tau}),$$

with E_2, E_4, E_6 the usual Eisenstein series given by

$$E_{2k}(\tau) = \frac{1}{2\zeta(2k)} \sum_{m,n} \frac{1}{(m\tau + n)^{2k}},$$

where the sum is over all integers m, n, not both 0, and τ is any complex number with positive imaginary part. Nesterenko proved that for any complex number q, with $0 < |q| < 1$, at least three of the numbers q, $P(q)$, $Q(q)$, $R(q)$ are algebraically independent over $\mathbb{Q}$. In particular, as he noted, on taking $q = e^{-2\pi}$ one has $P(q) = 3/\pi$, $Q(q) = 3(2\pi)^{-6}\Gamma(\frac{1}{4})^8$, $R(q) = 0$ and thus one obtains the striking result that π, e^π and $\Gamma(\frac{1}{4})$ are algebraically independent over $\mathbb{Q}$; even the algebraic independence of π and e^π was not known previously. Bertrand [39] applied Nesterenko's techniques to the study of theta-functions; he proved especially that the theta-function

$$\vartheta_3(z) = 1 + 2\sum_{n=1}^{\infty} z^{n^2}$$

is transcendental for $z = q$, where q is an algebraic number with $0 < |q| < 1$, and so in particular $\sum_{n=1}^{\infty} 2^{-n^2}$ is transcendental. This had hitherto been seen as an open problem that could not be readily handled by the Mahler method. Many further results on the values of theta and related functions are given in [187] and we refer there for details. Apart from the work of Barré-Sirieix *et al.*, Nesterenko's results rest on aspects of the Siegel–Shidlovsky theory as well as arguments from commutative algebra concerning multiplicity estimates on group varieties; for references to the latter see Chapter 5.

1.6 Riemann hypothesis over finite fields

Artin, motivated by classical work dating back to Gauss, conjectured in 1924 that the number N of solutions of the congruence $y^2 \equiv f(x) \pmod{p}$, where f denotes a cubic polynomial with integer coefficients and no multiple factors (mod p), satisfies

$$|N - p| \leq 2\sqrt{p}.$$

The conjecture was proved by Hasse in 1936 by way of studies on elliptic function fields and, from the standpoint of the theory of zeta-functions for elliptic curves, the result is analogous to the classical Riemann hypothesis. In his famous work of the 1940s, Weil obtained a far-reaching generalisation for non-singular projective curves of arbitrary genus g; namely he showed that the number N of F_q-rational points on such a curve, where F_q denotes the finite field with q elements, satisfies

$$|N - (q + 1)| \leq 2g\sqrt{q}.$$

Expressed affinely, this shows that the number N of solutions of the equation $f(x, y) = 0$, with f an absolutely irreducible polynomial over F_q, satisfies

$$|N - q| \leq 2g\sqrt{q} + c,$$

where c denotes a constant depending only on the degree of f. Weil's proof rested on deep work in algebraic geometry and it was the beginning of many important developments; in particular he conjectured an analogous result for any projective variety and this was proved in some profound work of Deligne.

In 1969 Stepanov discovered a new approach to questions of this kind. It was based on ideas from transcendence theory and did not involve a large background in algebraic geometry. The argument was subsequently developed by Bombieri [46] and by Schmidt [208], independently, to give a totally different proof of the Weil theorem; see these works for references to all the above.

Theorem 1.15 *The number N of solutions x, y in F_q of $f(x, y) = 0$ satisfies*

$$|N - q| \ll \sqrt{q},$$

where the implied constant depends only on the degree of f.

Though apparently weaker than the Weil theorem in view of the form of the constant, experts in the field will recognise that the results are equivalent and so Theorem 1.15 establishes the Riemann hypothesis for curves over finite fields. We shall give here a proof of the theorem in the case when the equation is given by $y^2 = f(x)$ where

$$f(x) = ax^3 + bx^2 + cx + d \quad (a \neq 0)$$

is an arbitrary cubic and F_q is the finite field with q elements with q an odd prime. Accordingly we shall show that the number N of solutions (x, y) of $y^2 = f(x)$ with integers x, y in F_q satisfies $|N - q| \ll \sqrt{q}$ where the implied constant is now absolute. We follow the exposition in [24] and, as we shall see, there are similarities with Sections 1.3 and 1.5. For another demonstration not involving algebraic geometry, due to Manin, see the book by Gelfond and Linnik [112].

Proof of Theorem 1.15. Let $g(x) = f(x)^{(q-1)/2}$ and let n, n' be the numbers of solutions of $g(x) = 1$ and $g(x) = -1$ respectively. By Fermat's little theorem, for every x in F_q we have either $f(x) = 0$ or $(f(x))^{q-1} = 1$. Since $g^2 = f^{q-1}$ we see that

$$n + n' \geq q - 3.$$

We proceed to prove that

$$\max\left(n, n'\right) \leq \tfrac{1}{2}q + O(\sqrt{q}),$$

where the constant implied in the O notation is absolute. Then we have $\left|n - \frac{1}{2}q\right| \ll \sqrt{q}$. But, by Euler's criterion, the solutions of $y^2 = f(x)$ are given by $(x, \pm y)$ for x such that $g(x) = 1$ and $(x, 0)$ for x such that $f(x) = 0$. Hence we see that $2n \leq N \leq 2n + 3$ and this gives

$|N - q| \ll \sqrt{q}$ as required. Here we are not concerned with the value of the implied constant and so we can assume $q > 5$.

We shall construct an auxiliary polynomial

$$\varphi(x) = \sum_{j=0}^{J-1} \big(p_{j0}(x) + p_{j1}(x)g(x)\big)x^{qj},$$

where the $p_{jk}(x)$ $(0 \le j < J,\ k = 0 \text{ or } 1)$ are polynomials, not all identically zero, with coefficients in F_q and with degrees at most $\frac{1}{2}(q-5)$. We show first that the terms in $\varphi(x)$, typically given by $p_{jk}(x)x^{qj}(g(x))^k$, have distinct degrees so that $\varphi(x)$ does not vanish identically. In fact, if d_{jk} is the degree of $p_{jk}(x)$, then the degree of the typical term is

$$d_{jk} + qj + \tfrac{3}{2}k(q-1).$$

Thus we have to verify that if

$$d_{jk} + qj + \tfrac{3}{2}k(q-1) = d_{j'k'} + qj' + \tfrac{3}{2}k'(q-1)$$

then $k = k'$ and $j = j'$. The equation gives

$$2(d_{jk} - d_{j'k'}) + 3(k' - k) = q\big(2(j' - j) + 3(k' - k)\big),$$

where both d_{jk} and $d_{j'k'}$ lie between 0 and $\frac{1}{2}(q-5)$. The assertion about distinct degrees now follows since k and k' are 0 or 1 whence the left-hand side has absolute value at most $q - 2$; but the right-hand side is a multiple of q and plainly non-zero unless $j = j'$ and $k = k'$.

Now we determine the $p_{jk}(x)$ such that

$$\varphi^{(l)}(x) = 0 \quad (0 \le l < L)$$

for all x with $g(x) = 1$ where $\varphi^{(l)}$ denotes the lth derivative of φ with respect to x. Plainly $\varphi^{(l)}$ has the same form as φ but with $p_{jk}(x)$ replaced by $p_{jkl}(x)/(f(x))^l$ where $p_{jkl}(x)$ denotes a polynomial with degree at most $\frac{1}{2}q + 3l$. Since $x^q = x$ for all x, it follows that one has to solve at most

$$(\tfrac{1}{2}q + 3L + J)\,L$$

linear equations in the $J(q-3)$ unknown coefficients of the $p_{jk}(x)$. This is possible if $L \ll \sqrt{q}$ with a sufficiently small constant and $J = \frac{1}{2}L + O(1)$. Note that as in the proof of Theorem 1.13 one does not need a Siegel lemma here. Finally one observes that $\varphi(x)$ has degree at most $(J+3)q$ and, since $L < q$, it has a zero at each solution of $g(x) = 1$ with order at least L. Hence we have

$$n \leq (J+3)q/L = \tfrac{1}{2}q + O(\sqrt{q})$$

provided that $L \gg \sqrt{q}$. Plainly n' can be treated similarly and this establishes our result. □

2
Logarithmic forms

2.1 Hilbert's seventh problem

In 1900, at the International Congress of Mathematicians held in Paris, Hilbert raised as the seventh of his famous list of problems the question whether $2^{\sqrt{2}}$ is transcendental and more generally whether α^β is transcendental for algebraic $\alpha \neq 0, 1$ and algebraic irrational β. He expressed the opinion that the solution lay farther in the future than the Riemann hypothesis or Fermat's last theorem.

Pólya [196] showed that among all transcendental entire functions which assume integer values for all non-negative integer values of the variable, that which increases the least is the function 2^z. The proof used an interpolation technique; for a generalisation to the function $2^{z_1+\cdots+z_n}$ of several variables see Baker [16]. Fukasawa [100] extended Pólya's work to Gaussian fields (for more recent results on this topic see Gramain [116]) and it was Gelfond's refinement [104] of this argument that originated his famous proof of the transcendence of $e^\pi = (-1)^{-i}$; this is a special case of Hilbert's seventh problem. In fact Gelfond proved that α^β is transcendental for algebraic $\alpha \neq 0, 1$ and imaginary quadratic β and Kuzmin [132] succeeded in extending the latter to real quadratic irrational β. But further progress awaited a new idea.

In 1934 Gelfond [105] and Schneider [211, I], independently, gave a complete answer to Hilbert's seventh problem. The work depended on the construction of an auxiliary function, a technique of the kind employed earlier by Siegel and Mahler (see Section 1.3 and Section 1.5) rather than direct appeal to $e^{\beta z}$ as in the Gelfond–Kuzmin work. In Gelfond's proof the function was a combination of α^z and $\alpha^{\beta z}$ and he used interpolation on the rational integers and differentiation; in Schneider's proof the function was a combination of z and α^z and he used extrapolation on the integer module generated by 1 and β and there was no differentiation. Now the Gelfond–Schneider theorem is a special case of a general theorem on meromorphic functions which we shall discuss in Section 2.3.

2.2 The Gelfond–Schneider theorem

We shall give here a proof of the Gelfond–Schneider theorem following essentially the method of Gelfond. Accordingly we shall prove that the ensuing holds.

Theorem 2.1 *Suppose that* $\alpha \neq 0, 1$ *and that* β *is irrational. Then* α, β *and* α^β *cannot all be algebraic.*

Proof. Suppose that α, β, α^β are algebraic. Let $\mathbb{K}$ be the number field $\mathbb{Q}(\alpha, \beta, \alpha^\beta)$ and suppose $\mathbb{K}$ has degree d. We denote by h, k positive integers, by $c_1, c_2, \ldots$ positive numbers which do not depend on h or k, and by $C_1, C_2, \ldots$ positive numbers which do not depend on k but can depend on h.

We construct an auxiliary function

$$\varphi(z) = \sum_{\lambda=0}^{L} \sum_{\mu=0}^{L} p(\lambda, \mu)\, \alpha^{(\lambda+\mu\beta)z},$$

where $L = \left[\sqrt{2dhk}\,\right]$ and the $p(\lambda, \mu)$ are integers, not all zero, such that

$$\varphi^{(j)}(l) = 0 \quad (1 \leq l \leq h, \quad 0 \leq j < k),$$

where $\varphi^{(j)}$ denotes the jth derivative of φ with respect to z. We have

$$\varphi^{(j)}(l) = (\log \alpha)^j \sum_{\lambda=0}^{L} \sum_{\mu=0}^{L} p(\lambda, \mu)\, (\lambda + \mu\beta)^j\, \alpha^{(\lambda+\mu\beta)l};$$

thus we have to solve $M' = hk$ linear equations in the $N = (L + 1)^2$ unknowns $p(\lambda, \mu)$. After multiplying by A^{2Lh+k} for a suitable integer A, a denominator for α, β, α^β, the coefficients in the equations become algebraic integers with sizes at most $(c_1 L)^k c_2^{Lh}$; here by the size of an algebraic integer we mean the maximum of the absolute values of the conjugates. Hence (see below) on expressing each coefficient in terms of an integral basis $\omega_1, \ldots, \omega_d$ of $\mathbb{K}$ we deduce that the M' linear equations with algebraic coefficients give rise to $M = dM'$ linear equations with integer coefficients, the latter having absolute values at most $(c_3L)^k c_2^{Lh}$. Hence by Siegel's lemma 1.11, since $N \geq c_4L^2$ and $M/(N-M) \leq 1$, the integers $p(\lambda, \mu)$ can be chosen to have absolute values $\leq (C_1k)^k$.

We proceed to prove by induction that if

$$\varphi^{(j)}(l) = 0 \quad (1 \leq l \leq h, \quad 0 \leq j < K)$$

with $K \geq k$ then $\varphi^{(K)}(l) = 0$ $(1 \leq l \leq h)$. We define

$$F(z) = (z-1)\cdots(z-h);$$

then $\varphi(z)/F(z)^K$ is an entire function and so by Cauchy's theorem

$$\frac{K!}{2\pi i}\int_{\mathcal{C}} \frac{\varphi(z)}{F(z)^K(z-l)}\,dz = \left[\frac{d^K}{dz^K}\left(\frac{\varphi(z)(z-l)^K}{(F(z))^K}\right)\right]_{z=l} = \frac{\varphi^{(K)}(l)}{(F'(l))^K},$$

where $\mathcal{C}$ is the positively orientated circle centred at the origin with radius $R = \sqrt{K}$ so that R exceeds $2h$ when $k \to \infty$. For z on $\mathcal{C}$ we have

$$|\varphi(z)| \leq (C_2K)^K c_6^{LR} \leq (C_3K)^K.$$

Further $|F(z)| \geq (R/2)^h$ and $|z-l| \geq R/2$ whence the expression on the left of the integral equation is

$$\leq C_4^K K!\, K^{(1-h/2)K}.$$

To estimate the right-hand side we observe that $(\log\alpha)^{-K}\varphi^{(K)}(l)$ is an algebraic number which after multiplying by A^{2Ll+K} becomes an algebraic integer with size (see Section 1.3) at most $(C_5K)^K$. Hence if $\varphi^{(K)}(l) \neq 0$ then, on taking norms, we get

$$\left|\varphi^{(K)}(l)\right| \geq (C_5K)^{-dK}.$$

Plainly if $h > 2(d+2)$ then since $\left|F'(l)\right| \leq C_6$ we have a contradiction for k and hence K sufficiently large. Thus $\varphi^{(K)}(l) = 0$ for $1 \leq l \leq h$.

We conclude by induction that $\varphi^{(k)}(l) = 0$ $(1 \leq l \leq h)$ for all k whence $\varphi(z)$ vanishes identically. In particular

$$\varphi^{(j)}(0) = 0 \quad \left(1 \leq j \leq (L+1)^2\right),$$

that is

$$\sum_{\lambda=0}^{L} \sum_{\mu=0}^{L} p(\lambda, \mu)\, (\lambda + \mu\beta)^j = 0 \quad \left(1 \leq j \leq (L+1)^2\right).$$

But the determinant of van der Monde type of order $(L+1)^2$ with terms $(\lambda + \mu\beta)^j$ is not zero since, by hypothesis, β is irrational and Theorem 2.1 follows. □

We now make some remarks on the details of the proof. First we recall that in order to apply Lemma 1.11, that is the basic form of the Siegel lemma, we used the fact that if an algebraic integer α in a field $\mathbb{K}$ has size A then it can be expressed as

$$\alpha = a_1\omega_1 + \cdots + a_d\omega_d\,,$$

where $\omega_1, \ldots, \omega_d$ is an integral basis for $\mathbb{K}$ and $a_1, \ldots, a_d$ are rational integers with absolute values $\ll A$. The latter estimate follows at once from the equation and its field conjugates which enable one to express each a_j as a linear combination of the conjugates of α. This indeed is the technique used in establishing a generalised version of Siegel's lemma to number fields [25, Ch. 5, §3] (similar to Lemma 1.12 but less sophisticated) and the conclusion could be used to simplify the proof of Theorem 2.1. Then we would need only the condition $N > M'$ and we could take $L = \left[\sqrt{2hk}\,\right]$ instead of $L = \left[\sqrt{2dhk}\,\right]$. In any event, our choice of L ensures that $L \ll \sqrt{k}$ whence $N = (L+1)^2 \ll M$ and $LR \ll k$. The latter is applied in the bound for the integral and it is a critical point in the proof.

One may be curious as to the origin of the integral equation. In fact it arises from the classical Newton interpolation formula. Let $\psi(z)$ be an entire function and let $\sigma_1, \ldots, \sigma_1, \ldots, \sigma_L, \ldots, \sigma_L$ be written as $\eta_1, \ldots, \eta_N$ where each σ is repeated K times so that $N = KL$. Then for any $w \in \mathbb{C}$ we have

$$\psi(w) = \sum_{n=0}^{N} a_n P_n(w),$$

where $P_0(w) = 1$, $P_n(w) = (w - \eta_1) \cdots (w - \eta_n)$ $(1 \leq n \leq N)$ and

$$a_n = \frac{1}{2\pi i} \int_C \frac{\psi(z)}{P_{n+1}(z)} \, dz \quad (0 \leq n < N),$$

$$a_N = \frac{1}{2\pi i} \int_C \frac{\psi(z)}{P_N(z)} \frac{dz}{(z - w)}$$

with C a circle centred at the origin that includes the η and w. To get the integral equation one defines the σ so that $P_N(z) = (F(z)/(z - l))^K$, one puts $\psi(w) = \varphi(w)/(w - l)^K$ and one takes limits as $w \to l$.

2.3 The Schneider–Lang theorem

Applications and developments of the Gelfond–Schneider theorem took two distinct tracks. Schneider was concerned mainly with elliptic and abelian functions; the principal results are now consequences of the Schneider–Lang theorem and its natural generalisation to functions of several variables. The theorem was developed by Schneider [213, 215] in the years 1934–1949 and given in the following form by Lang [137] in 1966.

Theorem 2.2 *Let $\mathbb{K}$ be an algebraic number field and let $f_1(z), \ldots, f_n(z)$ be meromorphic functions of finite order. Suppose that the ring $\mathbb{K}[f_1, \ldots, f_n]$ is mapped into itself by differentiation and has transcendence degree ≥ 2. Then there are only finitely many complex numbers z at which $f_1, \ldots, f_n$ simultaneously assume values in $\mathbb{K}$.*

A meromorphic function f is said to have finite order if, in the expression $f = g/h$ as a quotient of entire functions, we have for some fixed $\varrho > 0$

$$\max(|g(z)|, |h(z)|) < e^{R^{\varrho}}, \quad 0 \leq |z| \leq R \ (R > 1).$$

The transcendence degree of the ring $\mathbb{K}[f_1,\ldots,f_n]$ is the maximum number of elements in an algebraically independent subset. Theorem 2.1 follows at once from Theorem 2.2 on taking $f_1(z) = e^z, f_2(z) = e^{\beta z}$ and observing that, if α, β and α^β are algebraic, then f_1, f_2 simultaneously take values in $\mathbb{K} = \mathbb{Q}(\alpha, \beta, \alpha^\beta)$ when $z = l \log \alpha$ $(l = 1, 2, \ldots)$.

For a proof of Theorem 2.2 we refer to [25, Ch. 6]. One assumes that there exists a sequence of distinct complex numbers $y_1, y_2, \ldots$ such that $f_i(y_j)$ is an element of $\mathbb{K}$ for all i, j and one constructs by means of a version of Siegel's lemma an auxiliary function of the form

$$\Phi(z) = \sum_{\lambda_1=0}^{L} \sum_{\lambda_2=0}^{L} p(\lambda_1, \lambda_2)(f_1(z))^{\lambda_1}(f_2(z))^{\lambda_2}$$

such that $\Phi^{(j)}(y_l) = 0$ for all j, l with $0 \leq j \leq k$, $1 \leq l \leq m$. The argument then proceeds by induction with respect to j analogously to that of Theorem 2.1 using the integral equation

$$\frac{\Phi^{(j)}(y_l)}{(F'(y_l))^j} = \frac{j!}{2\pi i} \int_C \frac{\Phi(z)\,dz}{(z - y_l)(F(z))^j},$$

where $F(z) = (z - y_1) \cdots (z - y_m)$. We conclude that $\Phi^{(j)}(y_l) = 0$ $(1 \leq l \leq m)$ for all j whence Φ vanishes identically. Since the demonstration is valid with any pair from $f_1, \ldots, f_n$ in place of f_1, f_2, it follows that the transcendence degree of $\mathbb{K}[f_1, \ldots, f_n]$ is at most 1, contrary to hypothesis.

Schneider [213] showed in 1937 that if $\wp(z)$ is a Weierstrass $\wp$-function with algebraic invariants g_2, g_3 so that

$$\left(\wp'(z)\right)^2 = 4(\wp(z))^3 - g_2(\wp(z)) - g_3$$

then $\wp(\alpha)$ is transcendental for algebraic $\alpha \neq 0$. This now follows from Theorem 2.2 with $f_1(z) = \wp(\alpha z)$, $f_2(z) = \wp'(\alpha z)$, $f_3(z) = z$. As a corollary one deduces, by taking $\alpha = \frac{1}{2}\omega$, that any non-zero period ω of $\wp(z)$ is transcendental. In fact Schneider showed that $\alpha\omega + \beta\eta$ is transcendental where η is the quasi-period associated with a primitive ω, that is $\eta = 2\zeta\left(\frac{1}{2}\omega\right)$ where ζ is the Weierstrass ζ-function given by $\zeta'(z) = -\wp(z)$ and α, β are algebraic numbers not both zero. As he observed, this implies that the ellipse

$$\frac{x^2}{a^2} + \frac{y^2}{b^2} = 1$$

has a transcendental circumference if a, b are algebraic.

Another notable application of Theorem 2.2 due to Schneider is the transcendence of the elliptic modular function $j(z)$ where z is algebraic with positive imaginary part other than a quadratic irrational; in the exceptional case, $j(z)$ is algebraic with degree given by the class number of the quadratic field. For the proof we observe as in [25, Ch. 6, Theorem 6.3] that if α and $j(\alpha)$ are both algebraic then there is a $\wp$-function with algebraic invariants g_2, g_3 and fundamental periods ω_1, ω_2 such that $\alpha = \omega_2/\omega_1$. Now, by Theorem 2.2 applied to the functions $\wp(z)$, $\wp(\alpha z)$, $\wp'(z)$, $\wp'(\alpha z)$ which simultaneously take values in an algebraic number field when $z = (r + \frac{1}{2})\omega_1$ $(r = 1, 2, \ldots)$, we conclude that $\wp(z)$ and $\wp(\alpha z)$ are algebraically dependent whence α must be quadratic irrational. For a discussion giving generalisations to transcendence properties of automorphic functions see Section 8.5.

Theorem 2.2 has been extended to functions of several variables [214]. Here one uses partial derivatives mapping the ring into itself and the points at which the functions are assumed simultaneously to take values in the field are restricted to form a cartesian product of subsets of $\mathbb{C}$; in fact it suffices if the points do not lie on an algebraic hypersurface (see Bombieri and Lang [45, 49]). A striking theorem of Schneider in this context is the transcendence of the Beta-function

$$B(a,b) = \int_0^1 x^{a-1}(1-x)^{b-1}dx = \frac{\Gamma(a)\,\Gamma(b)}{\Gamma(a+b)}$$

for all rational, non-integral a, b (see [25, Ch. 6]). In the 1970s Chudnovsky [67] showed that $\Gamma(\frac{1}{3})$ and $\Gamma(\frac{1}{4})$ are transcendental and, as mentioned in Section 1.5, Nesterenko has recently developed this work to yield for instance the algebraic independence of π, $e^{\pi\sqrt{3}}$, $\Gamma(\frac{1}{3})$ and of π, e^{π}, $\Gamma(\frac{1}{4})$. These results represent all that is known to date on the transcendence of values of the Γ-function.

Gelfond followed another line of development arising from Theorem 2.1. He considered measures of transcendence and showed in 1935 [106] that if α and β are algebraic numbers such that the quotient $\log\alpha/\log\beta$ is irrational for some fixed branches of the logarithms then, for any $\varkappa > 5$,

$$\left|\frac{\log\alpha}{\log\beta} - \gamma\right| \gg e^{-(\log H)^{\varkappa}}$$

for all algebraic γ, where H denotes the height of γ (see Definition 1.5) and where the implied constant is effectively computable in terms of α, β, $\varkappa$ and the degree of γ. Gelfond [107] relaxed the condition $\varkappa > 5$ to $\varkappa > 3$ in 1939 and he further relaxed it in 1949 to $\varkappa > 2$ [109]. He also noted at about the same time that the work had some Diophantine applications. Thus for instance, having first extended the result just indicated to the p-adic domain, he proved that the equation

$$\alpha^x + \beta^y = \gamma^z$$

has only finitely many solutions in integers x, y, z if α, β, γ are real non-zero algebraic numbers, not all units or of the form $\pm 2^r$ with rational r. In his book [110] he remarked that if one could obtain a generalisation of his results concerning Diophantine approximation from two logarithms to arbitrarily many with similar effective estimations, then this would be of great consequence for the solution of many outstanding problems in number theory. He noted that from the Thue–Siegel theorem one can obtain an inequality of the form

$$|b_1 \log\alpha_1 + \cdots + b_n \log\alpha_n| \gg e^{-\delta B}$$

for algebraic $\alpha_1, \ldots, \alpha_n$ and rational integers $b_1, \ldots, b_n$ with absolute values at most B and for any $\delta > 0$. This improves upon the simple Liouville type estimate where δ is a fixed quantity that can be given explicitly in terms of the α. But the implied constant here, which depends on the α and δ, cannot be effectively computed, the reason being essentially the same as that discussed in Section 1.1. Thus the result had no direct bearing on problems of the kind in question and further progress awaited another approach.

2.4 Baker's theorem

A generalisation of the Gelfond–Schneider theorem to arbitrarily many logarithms was obtained by Baker [15, I,II] in 1966 and he extended the result in 1967 [15, III] to give the following.

Theorem 2.3 *If* $\alpha_1, \ldots, \alpha_n$ *are algebraic numbers, not* 0 *or* 1, *such that* $\log \alpha_1, \ldots, \log \alpha_n$ *are linearly independent over the rationals then* 1, $\log \alpha_1, \ldots, \log \alpha_n$ *are linearly independent over the field of all algebraic numbers.*

Here $\log \alpha_1, \ldots, \log \alpha_n$ are any fixed determinations of the logarithms. It is readily seen from Theorem 2.3 that any non-zero linear combination of $\log \alpha_1, \ldots, \log \alpha_n$ with algebraic coefficients is transcendental. Thus if we write

$$\Lambda = \beta_0 + \beta_1 \log \alpha_1 + \cdots + \beta_n \log \alpha_n ,$$

where the β are algebraic numbers, not all 0, then we have $\Lambda \neq 0$ if $\beta_0 \neq 0$ or if $\beta_0 = 0$ and $\beta_1, \ldots, \beta_n$ are linearly independent over the rationals. We refer to the conditions $\beta_0 = 0$ and $\beta_0 \neq 0$ as the homogeneous and inhomogeneous cases respectively. As an immediate corollary we have the following.

Theorem 2.4 *The number* $e^{\beta_0}\alpha_1^{\beta_1} \cdots \alpha_n^{\beta_n}$ *is transcendental for all non-zero algebraic* α *and* β. *Further, the number* $\alpha_1^{\beta_1} \cdots \alpha_n^{\beta_n}$ *is transcendental if* $1, \beta_1, \ldots, \beta_n$ *are linearly independent over* $\mathbb{Q}$.

Theorems 2.3 and 2.4 include as special cases the classical results of Hermite and Lindemann on the transcendence of e and π, as well as the Gelfond–Schneider theorem, and are plainly of interest from the point of view of the theory of transcendental numbers. However, of critical importance in connection with applications to the solution of Diophantine problems is the fact that the method of proof is effective and yields a sufficiently strong lower bound for $|\Lambda|$. After initial work in this context by Baker [15, III] yielding an estimate for $\Lambda \neq 0$ of the form $|\Lambda| \gg e^{-(\log B)^{\varkappa}}$ where $\varkappa > n + 1$, Feldman [95, 98] in 1971 improved the result to give the following theorem.

Theorem 2.5 *For any algebraic numbers* $\alpha_1, \ldots, \alpha_n$ *we have* $|\Lambda| \geq B^{-C}$ *for all algebraic numbers* $\beta_0, \beta_1, \ldots, \beta_n$ *with heights at* *most B (>1), where C is effectively computable in terms of the α and the degree of the β.*

The shape of the estimate, as far as it depends on B here, is best possible. Further, a great deal is now known about the form of the constant C; see the discussion in Section 2.8 and also Theorem 7.1. For proofs of Theorems 2.3, 2.4 and 2.5 see [25, Ch. 2 and 3]; we shall not repeat the demonstrations in detail here. However, to give some flavour of the theory we shall describe the proof of Theorem 2.5 in the so-called rational case when $\beta_0 = 0$, $\beta_1 = b_1, \ldots, \beta_n = b_n$ where $b_1, \ldots, b_n$ are rational integers, not all 0, and when $\alpha_1, \ldots, \alpha_n$ are multiplicatively independent, that is when $\alpha_1^{j_1} \cdots \alpha_n^{j_n} \neq 1$ if the exponents are integers not all 0. Accordingly we shall show that if $b_1, \ldots, b_n$ are rational integers, not all 0, with absolute values at most B (> 1) and if the linear form

$$\Lambda = b_1 \log \alpha_1 + \cdots + b_n \log \alpha_n$$

satisfies $|\Lambda| < B^{-C}$ for a sufficiently large constant C depending on $\alpha_1, \ldots, \alpha_n$ then the latter are multiplicatively dependent. In fact if C is sufficiently large then Λ vanishes, in other words if $\Lambda \neq 0$ then we have $|\Lambda| \geq B^{-C}$, but the proof is longer and we refer again to [25] for details.

2.5 The Δ-functions

The Δ-functions were introduced by Feldman [95] in order to obtain the version of Theorem 2.5 indicated above. He defined

$$\Delta(z; k) = \frac{(z+1) \cdots (z+k)}{k!}$$

for any positive integer k, and $\Delta(z; 0) = 1$. It is clear that $\Delta(z; k)$ takes integer values for all integers z and furthermore that, for any complex number z,

$$|\Delta(z; k)| \leq \frac{(|z| + k)^k}{k!} \leq e^{|z|+k}.$$

Soon afterwards Baker introduced the concept of generalised Δ-functions and this was crucial in his Sharpening Series of papers

[23] (see Section 2.8). Baker defined for any complex number z and integers $k \geq 0, l \geq 1, m \geq 0$

$$\Delta(z;k,l,m) = \frac{1}{m!}\frac{d^m}{dz^m}(\Delta(z;k))^l.$$

He proceeded to show that the functions have properties as described below.

Lemma 2.6 *We have*

$$|\Delta(z;k,l,m)| \leq e^{(|z|+2k)l}.$$

Proof. First we observe that

$$\Delta(z;k,l,m) = \Delta(z;k)^l \sum ((z+j_1)\cdots(z+j_m))^{-1},$$

where the sum is over all selections $j_1,\ldots,j_m$ from the set $1,\ldots,k$ repeated l times. Since

$$\Delta(z;k)^l((z+j_1)\cdots(z+j_m))^{-1} \leq \Delta(|z|;k)^l,$$

as we see by cancelling a factor and replacing it by another ≥ 1, and the number of terms in the sum is

$$\binom{kl}{m} \leq 2^{kl},$$

we obtain

$$|\Delta(z;k,l,m)| \leq 2^{kl}\Delta(|z|;k)^l.$$

The lemma now follows from the estimate for $\Delta(z;k)$ above. □

Lemma 2.7 *If $\nu(k)$ is the lowest common multiple of $1,\ldots,k$ then*

$$(\nu(k))^m \Delta(z;k,l,m)$$

is an integer for all integers z.

Proof. The result is due to Tijdeman [244] and it improves upon an earlier version in [25, Ch. 3, §2]. By a well known counting argument, the highest power of a prime p that divides $k!$ is given by $r = \sum_{j=1}^{s}\left[k/p^j\right]$

where p^s is the highest power of p that does not exceed k. The argument shows similarly that p^r divides $k!\Delta(z;k)$ (in fact this is clear since $\Delta(z;k)$ is an integer) and so p^{rl} divides $(k!\Delta(z;k))^l$. The lemma follows from the expression for $\Delta(z;k,l,m)$ in the proof of Lemma 2.6. For s is precisely the number of factors p of $\nu(k)$ and so, by the counting argument again, p^{rl} still divides $(k!\Delta(z;k))^l$ if we replace m of the factors by $\nu(k)$. □

Lemma 2.8 *We have* $\nu(k) \leq 4^k$.

Proof. The result follows at once from

$$\nu(k) = \prod_{p \leq k} p^{[\log k/\log p]} \leq k^{\pi(k)}$$

and the estimate $\pi(k) \leq (\log 4)k/\log k$ which is a consequence of refinements of the prime-number theorem; see [203]. □

Lemma 2.9 *The polynomials* $\Delta(z+\lambda';L'+1)^{\lambda+1}$, *where* $0 \leq \lambda' \leq L'$, $0 \leq \lambda \leq L$, *are linearly independent over* $\mathbb{Q}$.

Proof. The lemma follows from the independence of $1, x, \ldots, x^{n-m-1}$ and $P(x),\ P(x+1), \ldots, P(x+m)$ for any polynomial P defined over $\mathbb{Q}$ with degree $n > 0$ and any integer m with $0 \leq m \leq n$. The latter is proved in [25, Ch. 3, §3]. □

Lemma 2.10 *Let* $w_0, w_1, \ldots, w_{\sigma-1}$ *be distinct complex numbers. Using standard notation with the typical element in the* $(i+1)$*th row and* $(j+1)$*th column, the generalised van der Monde determinant* $\det\left(i^r w_s^i\right)$ *of order* $\varrho\sigma$, *where* $0 \leq i, j < \varrho\sigma$ *and* $j = r + \varrho s$ $(0 \leq r < \varrho, 0 \leq s < \sigma)$, *is not* 0.

Proof. See [25, Ch. 3, §3]. □

Lemma 2.11 *The* $(\tau+1)\sigma$ *by* $\tau\sigma$ *matrix with* $(i+1, j+1)$*th element*

$$\left(\Delta(i+\lambda', L'+1)\right)^{\lambda+1} w_s^i,$$

where $j = \lambda' + \lambda(L'+1) + s\tau$ *with* $0 \leq \lambda \leq L, 0 \leq \lambda' \leq L', 0 \leq s < \sigma$ *and* $\tau = (L'+1)(L+1)$, *has maximal rank* $\tau\sigma$.

Proof. If not then there exist elements $q(\lambda, \lambda', s)$, not all zero, such that

$$\sum_{\lambda=0}^{L}\sum_{\lambda'=0}^{L'}\sum_{s=0}^{\sigma-1} q(\lambda,\lambda',s)\Delta(i+\lambda';L'+1)^{\lambda+1}w_s^i = 0 \quad (0 \le i < (\tau+1)\sigma).$$

Now

$$\sum_{\lambda=0}^{L}\sum_{\lambda'=0}^{L'} q(\lambda,\lambda',s)\Delta(z+\lambda';L'+1)^{\lambda+1} = \sum_{r=0}^{\tau} q'(r,s)z^r,$$

where the $q'(r,s)$ are not all zero by virtue of Lemma 2.9. Hence

$$\sum_{r=0}^{\tau}\sum_{s=0}^{\sigma-1} q'(r,s)i^r w_s^i = 0 \quad (0 \le i < (\tau+1)\sigma)$$

with coefficients not all zero, which contradicts Lemma 2.10 with $\varrho = \tau + 1$. □

2.6 The auxiliary function

In the proof of Theorem 2.1 we constructed an auxiliary function of a complex variable z as a polynomial in α^z and $\alpha^{\beta z}$ with integer coefficients. Here we construct an auxiliary function in several complex variables $z_0, z_1, \ldots, z_{n-1}$ as a polynomial in $z_0, \alpha_1^{z_1}, \ldots, \alpha_{n-1}^{z_{n-1}}$ and

$$\alpha_1^{\beta_1 z_1} \cdots \alpha_{n-1}^{\beta_{n-1} z_{n-1}},$$

where $\beta_j = -b_j/b_n$ and, without loss of generality, it is assumed that $b_n \neq 0$.

To begin the proof of the result discussed in Section 2.4 we shall denote by $C_0, C_1, C_2, \ldots$ numbers that depend only on $\alpha_1, \ldots, \alpha_n$ and by k an integer that exceeds a sufficiently large C_0. Let $L = [k^{1-1/4n}]$ and $h = L' + 1 = [\log(kB)]$. The auxiliary function now takes the form

$$\varphi(z_0, z_1, \ldots, z_{n-1})$$
$$= \sum_{\lambda'=0}^{L'}\sum_{\lambda_0=0}^{L}\cdots\sum_{\lambda_n=0}^{L} p(\lambda)\Delta(z_0+\lambda';h)^{\lambda_0+1}\alpha_1^{\gamma_1 z_1}\cdots\alpha_{n-1}^{\gamma_{n-1}z_{n-1}},$$

where $\gamma_j = \lambda_j + \beta_j \lambda_n$ $(1 \leq j < n)$. The coefficients

$$p(\lambda) = p(\lambda', \lambda_0, \lambda_1, \ldots, \lambda_n)$$

will be determined as integers to satisfy

$$\sum_{\lambda'=0}^{L'} \sum_{\lambda_0=0}^{L} \cdots \sum_{\lambda_n=0}^{L} p(\lambda) \Delta(l + \lambda'; h, \lambda_0 + 1, m_0) \alpha_1^{\lambda_1 l} \cdots \alpha_n^{\lambda_n l} \gamma_1^{m_1} \cdots \gamma_{n-1}^{m_{n-1}} = 0. \tag{2.1}$$

Lemma 2.12 *There exist integers $p(\lambda)$, not all zero, such that $|p(\lambda)| \leq C_1^{hk}$ and (2.1) holds for all integers l with $0 \leq l < h$ and all non-negative integers $m_0, m_1, \ldots, m_{n-1}$ with*

$$m_0 + m_1 + \cdots + m_{n-1} < k .$$

Proof. Let A be a denominator for $\alpha_1, \ldots, \alpha_n$, that is a positive integer such that $A\alpha_1, \ldots, A\alpha_n$ are algebraic integers. Then, on multiplying the left-hand side of the equation (2.1) by $A^{nLl} b_n^k (\nu(h))^{m_0}$, we see that the coefficients of the $p(\lambda)$ become algebraic integers in the field $\mathbb{K} = \mathbb{Q}(\alpha_1, \ldots, \alpha_n)$ with sizes at most C_2^{hk}; this follows from Lemmas 2.6, 2.7 and 2.8 together with the inequality

$$|\gamma_j| \leq 2LB \leq e^{2h} \quad (1 \leq j < n),$$

which gives

$$|\gamma_1^{m_1} \cdots \gamma_{n-1}^{m_{n-1}}| \leq e^{2hk} .$$

We conclude that we have to solve $M' \leq hk^n$ linear equations in the $N = (L' + 1)(L + 1)^{n+1}$ unknowns $p(\lambda)$. Now since

$$(1 - 1/(4n))(n + 1) \geq n + 1/2$$

we have $N \geq hk^{n+1/2}$. Thus for $k > (2d)^2$, where d denotes the degree of $\mathbb{K}$, we obtain $N > 2M$ where $M = dM'$. It follows from Siegel's lemma, that is Lemma 1.11, that the equations have a non-trivial solution in integers $p(\lambda)$ with $|p(\lambda)| \leq C_1^{hk}$ as required. □

We now define $f(z) = f(z, m_0, \ldots, m_{n-1})$ as the function

$$\left(\frac{\partial}{\partial z_0}\right)^{m_0}\left(\frac{\partial}{\partial z_1}\right)^{m_1}\cdots\left(\frac{\partial}{\partial z_{n-1}}\right)^{m_{n-1}}\varphi(z_0, z_1, \ldots, z_{n-1})$$

evaluated at $z_0 = z_1 = \cdots = z_{n-1} = z$. The properties we shall require later are given by the following lemma.

Lemma 2.13 *For all non-negative integers $m_0, \ldots, m_{n-1}$ with $m_0 + \cdots + m_{n-1} < k$ we have*

$$|f(z)| < C_3^{hk+L|z|}.$$

Further, if l is an integer with $0 \leq l \leq hk^{8n}$, then either

$$|f(l)| < B^{-\frac{1}{2}C} \quad or \quad |f(l)| > C_4^{-hk-Ll}.$$

Proof. The first estimate follows at once from $|p(\lambda)| \leq C_1^{hk}$ together with Lemma 2.6 and straightforward bounds for the $\alpha_j^{\gamma_j z}$.

For the second part we note that if we replace the quantity

$$\alpha_n' = \alpha_1^{\beta_1} \cdots \alpha_{n-1}^{\beta_{n-1}}$$

which occurs in $f(l)$ by α_n so that the expression

$$\alpha_1^{\gamma_1 l} \cdots \alpha_{n-1}^{\gamma_{n-1} l} = \alpha_1^{\lambda_1 l} \cdots \alpha_{n-1}^{\lambda_{n-1} l} \alpha_n'^{\lambda_n l}$$

becomes $\alpha_1^{\lambda_1 l} \cdots \alpha_n^{\lambda_n l}$, then, apart from a factor

$$m_0!\,(\log \alpha_1)^{m_1} \cdots (\log \alpha_{n-1})^{m_{n-1}},$$

we obtain the left-hand side of (2.1). By the basic hypothesis on $|\Lambda|$ in Section 2.4 we see that

$$\left|\log \alpha_n - \log \alpha_n'\right| < B^{-C}$$

for some value of the second logarithm and some C which we suppose is sufficiently large in terms of k. Since $|e^z - 1| \leq |z|e^{|z|}$ for all complex numbers z, we obtain

$$\left|\alpha_n - \alpha_n'\right| < B^{-\frac{3}{4}C}.$$

We now consider l such that $0 \leq l \leq hk^{8n}$ and we observe that the left-hand side of (2.1) with such l, when multiplied by a denominator as before, becomes an algebraic integer with size at most C_5^{hk+Ll}. If it is zero then comparison with $f(l)$ and estimates similar to those used to obtain our bound for $|f(z)|$ give

$$|f(l)| < B^{-\frac{1}{2}C}.$$

If it is not zero then the norm is at least 1 and estimates for the conjugates together again with a comparison with $f(l)$ give $|f(l)| > C_4^{-hk-Ll}$. □

Note that we can cover the range $0 \leq l \leq hk^{8n}$ in Lemma 2.13 rather than $h < l \leq hk^{8n}$ as occurs in Lemma 5 in [25, Ch. 3, §3] since Lemma 2.7 above is sharper than the corresponding Lemma 1 in [25, Ch. 3, §2] and this eliminates a term $\log(l/h)$ from the lower bound given there for $|f(l)|$.

2.7 Extrapolation

In the proof of Theorem 2.1, the induction argument involved an extension in the order of derivation of the auxiliary function while keeping the h points of extrapolation fixed. Here we shall reduce the order of derivation $m_0 + \cdots + m_{n-1}$ but extend the points of extrapolation l. A key element in the exposition is the fact that the product of the number of relevant l with the number of relevant $m_0, \ldots, m_{n-1}$ increases in the inductive process.

Lemma 2.14 *For* $J = 0, 1, \ldots, (8n)^2$, *the number*

$$f(l) = f(l; m_0, \ldots, m_{n-1})$$

satisfies $|f(l)| < B^{-\frac{1}{2}C}$ *for all* $l, m_0, \ldots, m_{n-1}$ *with* $0 \leq l < hk^{J/(8n)}$ *and* $m_0 + \cdots + m_{n-1} < k/(2^J)$.

This will suffice to prove the desired result. For then the left-hand side of (2.1) vanishes for all l with $0 \leq l < (\tau + 1)\sigma$, where $\tau = (L' + 1)(L + 1)$ and $\sigma = (L + 1)^n$ (the N in the proof of Lemma 2.12 is then $\tau\sigma$), and $m_0 = \cdots = m_{n-1} = 0$. But from Lemma 2.11 we see that the matrix of coefficients of the $p(\lambda)$ in (2.1) has maximal rank $\tau\sigma$ unless

$$\alpha_1^{\lambda_1}\cdots\alpha_n^{\lambda_n} = \alpha_1^{\lambda_1'}\cdots\alpha_n^{\lambda_n'}$$

for some distinct vectors $(\lambda_1, \dots, \lambda_n)$ and $(\lambda_1', \dots, \lambda_n')$, that is unless $\alpha_1, \dots, \alpha_n$ are multiplicatively dependent.

Proof of Lemma 2.14. The lemma holds for $J = 0$ by Lemma 2.12 and the comparison argument used at the end of the proof of Lemma 2.13. We assume that it holds for $J = 0, 1, \dots, K$ and we proceed to prove the assertion for $J = K + 1$. Thus on defining

$$S_J = [hk^{J/(8n)}], \quad T_J = [k/2^J] \quad \text{for } J = 0, 1, \dots,$$

we have to prove that $|f(l)| < B^{-\frac{1}{2}C}$ for

$$S_K < l \le S_{K+1} \qquad \text{and} \qquad m_0 + \cdots + m_{n-1} < T_{K+1}.$$

Now let

$$F(z) = (z(z-1)\cdots(z-(S-1)))^T,$$

where for brevity we have written $S = S_K$, $T = T_{K+1}$. Then by Cauchy we have

$$\frac{1}{2\pi i}\int_C \frac{f(z)\,dz}{(z-l)F(z)} = \frac{f(l)}{F(l)} + \frac{1}{2\pi i}\sum_{s=0}^{S-1}\sum_{m=0}^{T-1}\frac{f_m(s)}{m!}\int_{C_s}\frac{(z-s)^m\,dz}{(z-l)F(z)},$$

where $S_K < l \le S_{K+1}$; here C is the positively orientated circle centred at the origin with radius $R = k^{1/(8n)}S_{K+1}$ and C_s is the circle centred at s with radius $\frac{1}{2}$. Further $f_m(z) = \left(\frac{d}{dz}\right)^m f(z)$, that is

$$\left(\frac{\partial}{\partial z_0} + \cdots + \frac{\partial}{\partial z_{n-1}}\right)^m \left(\frac{\partial}{\partial z_0}\right)^{m_0} \cdots \left(\frac{\partial}{\partial z_{n-1}}\right)^{m_{n-1}} \varphi(z_0, \dots, z_{n-1})$$

evaluated at $z_0 = \cdots = z_{n-1} = z$. Since $m_0 + \cdots + m_{n-1} < T = T_{K+1}$ and $0 \le m < T$ and furthermore $2T \le T_K$ we obtain from our inductive hypothesis

$$|f_m(s)| \le n^T B^{-\frac{1}{2}C} \le n^k B^{-\frac{1}{2}C} \quad (0 \le s < S,\ 0 \le m < T).$$

Clearly the integral over $\mathcal{C}_s$ above is bounded by $8^{ST}(S!)^{-T}$, and hence the double sum has absolute value at most $B^{-\frac{3}{8}C}(S!)^{-T}$ for C large enough. From Lemma 2.13 we see that, if we assume $|f(l)| \geq B^{-\frac{1}{2}C}$ contrary to the conclusion of Lemma 2.14 for $J = K+1$, then $|f(l)| > C_4^{-hk-Ll} > B^{-\frac{1}{8}C}$ and, since plainly $|F(l)| \leq 2^{S_{K+1}T}(S!)^T$, we obtain

$$|f(l)/F(l)| > B^{-\frac{1}{4}C}(S!)^{-T}.$$

It follows that the right-hand side of the integral equation has absolute value at least $\frac{1}{2}|f(l)/F(l)|$. This gives

$$|f(l)| \leq (4\vartheta/\Theta)|F(l)|,$$

where $\vartheta = \sup|f(z)|$ and $\Theta = \inf|F(z)|$ for z on $\mathcal{C}$. Now by Lemma 2.13 we have $\vartheta \leq C_3^{hk+LR}$ and clearly $\Theta \geq \left(\frac{1}{2}R\right)^{ST}$. Further we have $|F(l)| \leq (S_{K+1})^{ST}$ and thus, on observing that

$$LR \leq hk^{1+K/(8n)} \ll ST,$$

where the implied constant depends only on n, we obtain

$$|f(l)| < k^{-C_6 ST} \leq C_4^{-hk-Ll}.$$

But, by Lemma 2.13 again, this implies that $|f(l)| < B^{-\frac{1}{2}C}$ whence Lemma 2.14 holds by induction. □

2.8 State of the art

We shall be concerned in this section mainly with the rational case of the theory of logarithmic forms, that is

$$\Lambda = b_1 \log \alpha_1 + \cdots + b_n \log \alpha_n,$$

where $b_1, \ldots, b_n$ are integers. We shall assume that $\alpha_1, \ldots, \alpha_n$ are algebraic numbers with heights at most $A_1, \ldots, A_n$ (all $\geq e$) respectively and that the logarithms have their principal values. We assume further that

$b_1, \ldots, b_n$ have absolute values at most B ($\geq e$). The main result to date in the direction of Theorem 2.5 is due to Baker and Wüstholz [33].

Theorem 2.15 *If* $\Lambda \neq 0$ *then*

$$\log|\Lambda| > -(16nd)^{2(n+2)} \log A_1 \cdots \log A_n \log B,$$

where d *denotes the degree of* $\mathbb{Q}(\alpha_1, \ldots, \alpha_n)$.

The theorem is best possible with respect to $A_1, \ldots, A_n$ and B separately and moreover the function of n and d involves only small numerical constants and is quite sharp. These features are important in applications as we shall see from the next chapter. In fact Baker and Wüstholz give a still stronger and more refined version in terms of the logarithmic Weil heights of the α; this will be discussed in Section 7.2. There has been an improvement relating to the expression involving n and d due to Matveev which we shall refer to again later, but the basic structure of the work remains the same. It is still essentially the best result of its kind.

Here we shall describe a little of the history. If we follow the proof of Theorem 2.5 we obtain a result of the form if $\Lambda \neq 0$ then

$$|\Lambda| > \exp\left(-C(\log A)^{\varkappa} \log B\right),$$

where A denotes the maximum of the heights of the α; here $\varkappa$ depends on n and C depends on n and d. The result holds for the general linear form Λ as in Theorem 2.5, with B given by the maximum of the heights of $\beta_0, \beta_1, \ldots, \beta_n$. The original value for $\varkappa$ was quite large, in fact of order at least n^2 (see Baker [15, IV]), but stronger results could be obtained if one of the α, say α_n, had a large height relative to the remainder; this frequently occurs in applications. Thus it was shown that if A is the height of α_n and if A' is a bound for the heights of $\alpha_1, \ldots, \alpha_{n-1}$ then the above inequality is valid for any $\varkappa > n$ provided that C is allowed to depend on A' as well as on n and d (see Baker [17, I]). The condition was relaxed to $\varkappa > n-1$ by Feldman [96, 97] in 1969. It was further relaxed to $\varkappa > 1$ by Baker and Stark [31] in 1971. The latter work was motivated by the class number two problem (see Section 3.1) and the argument involved the introduction of Kummer theory; this has played an important role in all subsequent studies. In 1972, Baker [23, I] (the papers in [23] have come to be known as the 'Sharpening Series') reduced the condition to $\varkappa = 1$ which is best possible; the main innovation here was the introduction of the generalised Δ-functions as discussed in Section 2.5.

The question now arose as to how C depends on A'; this was important to Tijdeman [244] in his work on the Catalan conjecture (see Section 3.6). He calculated a value for C of the form $C'(\log A')^{\varkappa'}$ where again $\varkappa'$ depends on n and where C' depends on n and d. In 1975 by a new reduction technique, the 'Kummer descent', Baker [23, III] derived the result

$$|\Lambda| > \exp(-C'\Omega \log \Omega \log B),$$

where $\Omega = \log A_1 \cdots \log A_n$; it was here that an estimate in the shape of Theorem 2.15 first appeared. Shortly afterwards van der Poorten [197] noted that $\log \Omega$ could be replaced by $\log \Omega'$ where $\Omega' = \Omega / \log A_n$; then the inequality included all previously quoted results as special cases.

The next question that arose was what form does C' take as a function of n and d? Some studies in this context had already been carried out in connection with the early inequalities in the subject and they had yielded expressions with $\log C'$ of order n^4 (see Baker [15, IV]). The motivation amongst other things was applications to classical theories furnishing estimates for the greatest prime factor of polynomials and of binary forms. After improvements by Sprindžuk and Kotov in this context (for references see [234]) the exponent was reduced, in 1975, to one of order $n \log n$ by Shorey [224]. The latter work involved a new idea concerning the size of the inductive steps (cf. Lemma 2.14) and it established the basic form of the expression for n. In 1977, by a combination of the preceding techniques, Baker [26] gave the result

$$|\Lambda| > \exp\left(-(16nd)^{200n}\,\Omega \log \Omega' \log B\right).$$

Further, he obtained the same inequality in the general case, that is with algebraic β, with B replaced by $B\Omega$.

The main problem over the decade beginning 1977 was to eliminate the $\log \Omega'$ term; this was successfully solved by Wüstholz [262] and by Philippon and Waldschmidt [194], independently, using the theory of multiplicity estimates on group varieties. This is now an important instrument in transcendence theory and its discovery and development has been a major achievement; see the discussion beginning in Chapter 4. The theory of multiplicity estimates has also led to improvements in the constants occurring in the expression for n and the best form of the result is now given by Theorem 7.1. A result of a similar kind was proved at about the same time by Waldschmidt [252] but this did not involve a Kummer descent and Theorem 7.1 is consequently stronger.

Recently some work of Matveev [174] has appeared which gives an improved form for the expression for n of the shape c^n for an absolute constant c. Matveev's articles contain a number of new elements and they constitute an important advance; for a discussion in the simplest case see the article by Nesterenko in [3, pp. 53–106] and for further remarks see Section 7.2.

Theorem 2.15 is capable of generalisation in several directions. First one would expect a similar result for the linear form Λ with algebraic coefficients, taking $B\Omega$ in place of B, in other words one would expect a result as in Baker [26], but the details have not been given as yet. It is worth observing that as a very special instance of [26] we have the best measure of irrationality for e^π established to date, namely

$$\left|e^\pi - p/q\right| > q^{-c \log\log q},$$

valid for all rational p/q, where c is an absolute constant. The exponent $\log\log q$ arises from the extra Ω factor attaching to B; conjecturally one would expect $|e^\pi - p/q| \gg q^{-2-\varepsilon}$ for any $\varepsilon > 0$ but this would involve the deletion of the extra factor and this seems to present considerable difficulty.

Another area of generalisation of Theorem 2.15, of particular interest in connection with Diophantine studies, is in the direction of Baker's Sharpening II [23]. Subject to slight modifications in the constants occurring in the expression for n, the factor $\log B$ in the estimate for $|\Lambda|$ can be replaced by $\log B'$ where

$$B' = \max_{1 \le j < n} \left\{ \frac{|b_n|}{\log A_j} + \frac{|b_j|}{\log A_n} \right\},$$

and it is assumed that $b_n \neq 0$. A verification of a result of this kind follows easily from the exposition of Baker and Wüstholz [33] and details of such a deduction can be found in the paper of Waldschmidt [252]; for further discussion in this context see Section 7.2.

In the case of linear forms in two logarithms there is a variant of the basic theory based on interpolation determinants rather than the box principle. The method was originated by Laurent [143] and it yields lower bounds of the form

$$\log|\Lambda| > -cd^4 \log A_1 \log A_2 (\log B)^2,$$

where the notation is that of Theorem 2.15. Though plainly weaker than the latter theorem in regard to the dependence on B, it furnishes relatively small numerical values for the constant c and this is important in applications. For an account of the topic we refer to the book of Waldschmidt [253].

Finally we mention that there is an extensive theory generalising studies on logarithmic forms to the p-adic domain. The theory has a long history following closely the results in the complex domain. Thus in 1935 Mahler [154] obtained a p-adic analogue of the Gelfond–Schneider theorem; Coates [70] and independently Sprindžuk [233] (see also Brumer [62] and Vinogradov and Sprindžuk [248]) obtained p-adic analogues of Baker's first results on logarithmic forms, and van der Poorten [198] gave p-adic analogues of Baker's 1977 results [26]. Subsequently, substantial papers by Kunrui Yu [267] were published which overcame certain inaccuracies that he detected in van der Poorten's exposition and significantly improved the numerical estimates. More recently Kunrui Yu [268] has worked out p-adic analogues of Baker and Wüstholz [33] and this area of research is continuing at the present time; see [269]. These works have been of great importance in connection with theories relating to p-adic L-functions, Diophantine geometry and elsewhere, and we shall discuss applications of the theory in the next chapter.

3
Diophantine problems

3.1 Class numbers

Gauss conjectured that the only imaginary quadratic fields $\mathbb{Q}(\sqrt{-d})$ with class number 1 are given by $d = 1, 2, 3, 7, 11, 19, 43, 67$ and 163. This was solved as one of the first applications of the theory of linear forms in logarithms; see Baker [15, I]. The approach arose from studies of Gelfond and Linnik [111]. Another solution motivated by earlier work of Heegner [126] on elliptic modular functions was given by Stark [235] at about the same time. Both depend on an analogue of the classical Kronecker limit formula (see [22] and [25, Ch. 5]).

Let $-d < 0$ and $k > 0$ denote the discriminants of the quadratic fields $\mathbb{Q}(\sqrt{-d})$ and $\mathbb{Q}(\sqrt{k})$ respectively. Suppose that $(k, d) = 1$. Let $\chi(n) = (\frac{k}{n})$ and let $\chi'(n) = (\frac{-d}{n})$ be characters given by the usual Kronecker symbols. Then for any $s > 1$ we have

$$L(s, \chi)\, L(s, \chi\chi') = \frac{1}{2} \sum_{f} \sum_{x,y} \chi(f) f^{-s},$$

where x, y run through all integers not both zero, and

$$f = f(x, y) = ax^2 + bxy + cy^2$$

runs through a complete set of inequivalent quadratic forms with discriminant $-d$; on the left we have the usual L-functions given by

$$L(s, \chi) = \sum_{n} \chi(n)\, n^{-s}.$$

On taking limits as $s \to 1$ we obtain

$$L(1,\chi)\,L(1,\chi\chi') = \frac{1}{6}\pi^2 \prod_{p|k}\left(1-\frac{1}{p^2}\right)\sum_f \frac{\chi(a)}{a} + \sum_f \sum_{r=-\infty}^{\infty} A_r e^{\pi i r b/(ka)},$$

where, for $r \neq 0$, we have $|A_r| \leq 2r' e^{-r'/(ka)}$ with $r' = \pi|r|/\sqrt{d}$, and $A_0 = 0$ unless k is the power of a prime p, in which case

$$A_0 = \frac{-2\pi}{k\sqrt{d}}\chi(a)\log p.$$

Suppose now that $\mathbb{Q}(\sqrt{-d})$ has class number 1. Then, by the theory of genera, d is prime and $\equiv 3 \pmod 4$ assuming $d > 2$. Further there is just one form f which can be taken as the principal form

$$x^2 + xy + \frac{1}{4}(1+d)y^2.$$

By classical results of Dirichlet we have

$$L(1,\chi) = \frac{2h(k)}{\sqrt{k}}\log \varepsilon_k, \quad L(1,\chi\chi') = h'(k)\frac{\pi}{\sqrt{kd}},$$

where $h(k)$, $h'(k)$ denote the class numbers of $\mathbb{Q}(\sqrt{k})$ and $\mathbb{Q}(\sqrt{-kd})$ respectively and ε_k is the fundamental unit in $\mathbb{Q}(\sqrt{k})$. Thus assuming that $d > k$, so that $(d,k) = 1$, we obtain

$$\left|2\,h(k)\,h'(k)\,\log\varepsilon_k - \frac{1}{6}\pi\,k\sqrt{d}\prod_{p|k}\left(1-\frac{1}{p^2}\right)\right| \leq k\frac{\sqrt{d}}{\pi}\sum_{r=-\infty}^{\infty}|A_r|.$$

Further, if k is not a power of a prime p then $A_0 = 0$ and the estimate for $|A_r|$ then implies that the expression on the right is at most

$$16k\,e^{-\pi\sqrt{d}/k}$$

assuming $\sqrt{d}/k > 1$. We apply the inequality with $k = 21$ and $k = 33$. In both cases $h(k) = 1$ and for sufficiently large d ($d > 10^{20}$) we get

$$\left|h'(21)\log\varepsilon_{21} - \frac{32}{21}\pi\sqrt{d}\right| < e^{-\pi\sqrt{d}/100}$$

and similarly with 33 in place of 21 and 80 in place of 32. On eliminating $\pi\sqrt{d}$ it follows that

$$\left|35\,h'(21)\log\varepsilon_{21} - 22\,h'(33)\log\varepsilon_{33}\right| < 57\,e^{-\pi\sqrt{d}/100}.$$

This can be expressed in the form

$$|b_1 \log\alpha_1 + b_2 \log\alpha_2| < e^{-\delta B}$$

with the notation of Section 2.4 and since $\alpha_1 = \varepsilon_{21}$ and $\alpha_2 = \varepsilon_{33}$ are multiplicatively independent we deduce from Theorem 2.5 that the left-hand side is greater than d^{-C} for some numerical constant C. It follows that d is bounded, that is there are only finitely many imaginary quadratic fields with class number 1. In practice we obtain an explicit bound for d (originally 10^{500}, now about 10^{20}) and it is then easy to show by computing the continued fraction for $\log\alpha_1/\log\alpha_2$ that there are only the nine cases listed by Gauss.

A similar argument applies for the determination of all imaginary quadratic fields with class number 2; in this case the problem reduces to an inequality involving three logarithms and one needs a sharp version of the linear form result as described in Section 2.8. It has thus been established that there are precisely eighteen fields in question (see Baker [21]; Stark [236]).

In 1976 Goldfeld [114] found a new approach to the topic based on the theory of elliptic curves and Gross and Zagier [121] succeeded in this way to show that the class number $h(d)$ of $\mathbb{Q}(\sqrt{-d})$ satisfies

$$h(d) \gg (\log d)^{1-\varepsilon}$$

for any $\varepsilon > 0$ where the implied constant depends only on ε and is effectively computable; thus all imaginary quadratic fields with any given class number can now be determined in principle. A classical theorem of Siegel [230] asserts that

$$h(d) \gg d^{\frac{1}{2}-\varepsilon}$$

and the exponent here is best possible; but the implied constant cannot be effectively computed in view of the well-known question of the non-existence of the Siegel zero for Dirichlet L-functions. All the imaginary quadratic fields $\mathbb{Q}(\sqrt{-d})$ with class number $h(d) = 3$ have now been calculated by Oesterlé, with $h(d) = 4$ by Arno, with $h(d) = 5, 6, 7$ by Wagner and with all odd $h(d) \leq 23$ by Arno, Robinson and Wheeler [7]. Recently Watkins has published a paper [254] which purports to give a complete determination of all the imaginary quadratic fields with $h(d) \leq 100$; he says that the work took about seven months of computation. We refer to [7] and [254] for references and to Goldfeld [115] for further details about the class number problem.

The theory of linear forms in logarithms also shows that the numbers $L(1, \chi)$ taken over all non-principal characters χ with prime modulus q are linearly independent over $\mathbb{Q}$; this generalises Dirichlet's famous result on the non-vanishing of $L(1, \chi)$ (see Baker, Birch and Wirsing [35]). Furthermore it shows that the corresponding value $L_p(1, \chi)$ of the p-adic L-function is not 0 for any non-principal character χ. This was originally a conjecture of Leopoldt; the result follows from ideas of Ax [10] together with an estimate for linear forms in p-adic logarithms. A particular result sufficient for this purpose was first given by Brumer [62] by straightforward adaptation of Baker's original work [15] and there have been considerable developments in the p-adic theory since then; see Section 2.8. The problem of proving the non-vanishing of the p-adic regulator of a non-abelian field remains open.

3.2 The unit equations

Originally implicit in the study of Diophantine equations, the unit equations now form a major topic with numerous applications.

Let $\mathbb{K}$ be an algebraic number field with degree d and with ring of integers $\mathcal{O}_{\mathbb{K}}$. Let S be a finite set of places of $\mathbb{K}$ including all the infinite places of $\mathbb{K}$. An element α of $\mathbb{K}$ is said to be an S-unit if

$$|\alpha|_v = 1$$

for each place v not in S, where $|\cdot|_v$ denotes a suitably normalised valuation such that the product formula holds. The S-units form a finitely generated multiplicative group U_S; if S contains no finite places then U_S is just the group of units $U_{\mathbb{K}}$ of $\mathcal{O}_{\mathbb{K}}$.

We consider the inhomogeneous S-unit equation in two variables, namely

$$\alpha x + \beta y = 1,$$

where α, β are non-zero elements of $\mathbb{K}$. The basic result is as follows.

Theorem 3.1 *There are only finitely many solutions of the equation in S-units x and y and all of these can be effectively determined.*

Proof. We shall give the details in the case when U_S is $U_{\mathbb{K}}$ and indicate later how one derives the more general result. Let $\eta_1, \ldots, \eta_r$ be the $r = s + t - 1$ fundamental units in $\mathbb{K}$ given by Dirichlet's theorem. Then we have

$$x = \varrho\, \eta_1^{x_1} \cdots \eta_r^{x_r}, \quad y = \sigma\, \eta_1^{y_1} \cdots \eta_r^{y_r}$$

for some rational integers $x_1, \ldots, x_r$, $y_1, \ldots, y_r$ and some roots of unity ϱ, σ. We write $X = \max |x_j|$ and $Y = \max |y_j|$. Our object is to show that $\max(X, Y)$ is bounded. We suppose without loss of generality that $Y \geq X$. We first establish the theorem on the assumption that

$$\log |y| \gg Y,$$

where the implied constant depends only on $\mathbb{K}$; this will not hold in general but, as we shall verify in a moment, it is valid for some conjugate y' of y and, as will be clear, this suffices for our purpose.

The equation

$$\alpha x + \beta y = 1$$

gives

$$\log |(\alpha x/\beta y) + 1| = -\log |\beta y| \ll -Y$$

assuming that $Y \gg 1$. Since, for any complex number z, the inequality

$$\left|e^z + 1\right| \leq \tfrac{1}{4}$$

implies that

$$|z - ik\pi| \leq 4\left|e^z + 1\right|$$

for some rational integer k, we obtain

$$\log |\Lambda| \ll -Y,$$

where $\Lambda = \log(\alpha x/\beta y) - k \log(-1)$. But

$$\log(\alpha x/\beta y) = \log(\alpha\varrho/\beta\sigma) + \sum_{j=1}^{r}(x_j - y_j)\log\eta_j,$$

where the logarithms on the right can be assumed to have their principal values. Furthermore we have $|x_j - y_j| \leq 2Y$ and this gives $|k| \ll Y$. It is readily seen that $\Lambda \neq 0$ whence Theorem 2.5 implies that

$$\log|\Lambda| \gg -\log Y.$$

Thus $Y \ll \log Y$ and so Y is bounded. Hence also X is bounded and so there are only finitely many x, y as asserted.

Plainly we can begin with an equation conjugate to $\alpha x + \beta y = 1$ and thus it remains only to prove that

$$\log|y'| \gg Y$$

for some conjugate y' of y. Accordingly let $\alpha^{(1)}, \ldots, \alpha^{(d)}$ be the conjugates of any element α of $\mathbb{K}$ with the usual ordering with respect to real and complex conjugates. Then

$$\det\left(\log\left|\eta_i^{(j)}\right|\right) \quad (1 \leq i, j \leq r)$$

is the regulator R of $\mathbb{K}$ and we have $R \neq 0$. Hence $y_1, \ldots, y_r$ can be expressed as linear combinations of $\log|y^{(1)}|, \ldots, \log|y^{(r)}|$ with coefficients given by minors of order $(r-1)$ of R. This gives

$$Y \ll \max\left|\log\left|y^{(j)}\right|\right|.$$

Let the maximum be given by $j = l$; then either $\log|y^{(l)}| \gg Y$ or $\log|y^{(l)}| \ll -Y$. In the first case we have the desired assertion. In the second case we recall that y is a unit and thus

$$\sum_{j=1}^{d}\log\left|y^{(j)}\right| = 0.$$

Hence we have $\log|y'| \gg Y$ for some conjugate y' of y as asserted. □

To establish Theorem 3.1 for the general system of S-units U_S we denote by $\mathfrak{p}_1, \ldots, \mathfrak{p}_s$ the prime ideals corresponding to the finite places of S. Then $\mathfrak{p}_1^h, \ldots, \mathfrak{p}_s^h$, where h is the class number of $\mathbb{K}$, are principal ideals. Suppose they are generated by elements $\pi_1, \ldots, \pi_s$ in $\mathcal{O}_{\mathbb{K}}$. Every element of U_S can be expressed in the form

$$\gamma \eta_1^{u_1} \cdots \eta_r^{u_r} \pi_1^{v_1} \cdots \pi_s^{v_s}$$

for some rational integers $u_1, \ldots, u_r$, $v_1, \ldots, v_s$ and some γ in $\mathcal{O}_{\mathbb{K}}$ belonging to a finite set. This holds in particular for the solutions x, y of the S-unit equation and Theorem 3.1 follows by first applying the p-adic theory of linear forms in logarithms and then the classical theory. The work leads to an explicit upper bound for the sizes of the solutions; thus for example the rational case amounts to an effective solution of the equation

$$a p_1^{x_1} \cdots p_t^{x_t} + b p_1^{y_1} \cdots p_t^{y_t} = c$$

with positive integers a, b, c. A bound for the absolute values of the x and y in terms of $\max(a, b, c)$ and $\max(p_1, \ldots, p_t)$ has been given by Győry. For extensive references in this connection see the excellent survey [123].

It should be mentioned that, by a p-adic version of Schmidt's subspace theorem [209, 210] which we shall discuss in Chapter 8, one can show that the equation

$$\alpha_1 x_1 + \cdots + \alpha_n x_n = 1,$$

where $\alpha_1, \ldots, \alpha_n$ are non-zero elements of $\mathbb{K}$, has only finitely many solutions in S-units $x_1, \ldots, x_n$ provided that no proper subsum of $\alpha_1 x_1 + \cdots + \alpha_n x_n$ vanishes; see the survey article by Evertse *et al.* [88]. The argument is ineffective in the sense that it does not allow one to determine the complete list of solutions in any given instance (cf. Section 1.1) but there is extensive work by Schlickewei, Evertse and others furnishing explicit estimates for the number of solutions; see, in particular, [86, 87].

3.3 The Thue equation

Studies on Diophantine equations by way of techniques in transcendence theory were originated by Thue [243] in 1909. Let

$$F(x, y) = a\,(x - \alpha_1 y) \cdots (x - \alpha_n y)$$

be an irreducible binary form with integer coefficients and degree $n \geq 3$. Thue's famous theorem, which is given as follows, can now be demonstrated directly from Theorem 3.1 and indeed effectively.

Theorem 3.2 *The equation $F(x, y) = m$, where m is any integer, has only finitely many solutions in integers x and y.*

Proof. Let $\mathbb{K}$ be the algebraic number field generated by $\alpha_1, \dots, \alpha_n$ over $\mathbb{Q}$. Since there are only finitely many non-associated elements of $\mathcal{O}_{\mathbb{K}}$ with a given norm, we have

$$x - \alpha_j y = \gamma_j \eta_j \quad (1 \leq j \leq n),$$

where $\eta_1, \dots, \eta_n$ are units in $\mathbb{K}$ and $\gamma_1, \dots, \gamma_n$ belong to a finite effectively computable set. Now the identity

$$(\alpha_3 - \alpha_2)(x - \alpha_1 y) + (\alpha_1 - \alpha_3)(x - \alpha_2 y) + (\alpha_2 - \alpha_1)(x - \alpha_3 y) = 0$$

gives

$$\gamma_1' \eta_1 + \gamma_2' \eta_2 + \gamma_3' \eta_3 = 0$$

with obvious definitions for γ_1', γ_2', γ_3'; these are non-zero if we assume, as we may, that α_1, α_2, α_3 are distinct. Hence $\alpha x' + \beta y' = 1$ with $\alpha = -\gamma_1'/\gamma_3'$ and $\beta = -\gamma_2'/\gamma_3'$ and with x', y' given by the units η_1/η_3 and η_2/η_3. From Theorem 3.1 we see that there are only finitely many x', y' and thus also only finitely many rationals x/y. This establishes the result. □

The work leads to an explicit bound for $\max(|x|, |y|)$ of the form

$$(C_1 |m|)^{C_2},$$

where C_1, C_2 depend only on F. The first result of this kind was given by Baker [17, I] in 1968 but with a weaker dependence on m; the form given here, which is best possible in terms of m, is a consequence of a refinement of Theorem 2.5 relating to the case when $b_n = 1$; it is discussed in a more advanced setting at the end of Section 7.2. A result of the kind was obtained explicitly in Baker's Sharpening II [23] and implicitly in Feldman [98]. Expressions for C_1, C_2 can be given in terms of n, the coefficients of F and the regulator of $\mathbb{K}$; the best results to date are due to Bugeaud and Győry and we refer to [123] for details.

The estimate above for the solutions of the Thue equation yields at once an effective improvement on Liouville's theorem; see Section 1.1. Indeed, if $\alpha = \alpha_1$ and if p/q, $q > 0$, is any rational with $|\alpha - p/q| < 1$, then $|F(p,q)| \ll q^n|\alpha - p/q|$ where the implied constant depends only on F. But we have $|F(p,q)| \gg q^\delta$, where $\delta = 1/C_2 > 0$, and so

$$|\alpha - p/q| \gg q^{-\varkappa}$$

with $\varkappa = n-\delta$. The value of δ here depends on the height of α as well as on the degree n; for some explicit numerical examples relating to cubic irrationals see Baker and Stewart [32].

3.4 Diophantine curves

The equation $y^2 = x^3 + k$ has a long history dating back to Mordell in the early 1900s. After initially conjecturing that for certain specified k there would be infinitely many solutions in integers x, y while for others there would be only finitely many, Mordell became aware of the work of Thue and proved that in fact for any integer $k \neq 0$ the equation has only finitely many solutions. The problem of actually determining the complete list of solutions for a specified value of k was subsequently a focus of much attention and various ad hoc methods were devised to deal with particular instances [182].

As an early application of the theory of logarithmic forms, Baker [17, II] gave a new and effective proof of the finiteness of the number of the solutions of the Mordell equation and, shortly afterwards, he extended the result to the hyperelliptic equation $y^2 = f(x)$ where

$$f(x) = a(x - \alpha_1)\cdots(x - \alpha_n)$$

denotes a polynomial with integer coefficients and it is assumed that $\alpha_1, \alpha_2, \alpha_3$ are simple zeros of f. A result of the same kind based on the Thue–Siegel theory was proved by Siegel [227] in 1926 but for the reason indicated in Section 1.1 the conclusion is ineffective. Here we shall show, following [227] and Baker [19], that the result can be deduced readily and effectively from Theorem 3.1, that is from the unit theorem.

Theorem 3.3 *There are only finitely many solutions to the equation $y^2 = f(x)$ in integers x, y and these can be effectively determined.*

Proof. Let $\mathbb{K}$ be the field generated by the α over $\mathbb{Q}$. Suppose that x, y is a solution of $y^2 = f(x)$ and let $\beta_j = x - \alpha_j$ $(1 \leq j \leq n)$ so that $f(x) = a\beta_1 \cdots \beta_n$. Then since, by assumption, α_1, α_2, α_3 are simple zeros, we have, by considering ideals in $\mathbb{K}$,

$$\beta_j = \gamma_j \zeta_j^2 \quad (j = 1, 2, 3),$$

where γ_j, ζ_j are in $\mathbb{K}$ and the γ_j belong to a finite set.

Now consider the identity

$$\left(\sqrt{\beta_3} - \sqrt{\beta_2}\right) + \left(\sqrt{\beta_1} - \sqrt{\beta_3}\right) + \left(\sqrt{\beta_2} - \sqrt{\beta_1}\right) = 0.$$

Each number in parentheses belongs to the field

$$\mathbb{K}' = \mathbb{K}\left(\sqrt{\gamma_1}, \sqrt{\gamma_2}, \sqrt{\gamma_3}\right)$$

and has a fixed norm independent of x and y; for instance, in the typical case,

$$N_{\mathbb{K}'}\left(\sqrt{\beta_3} - \sqrt{\beta_2}\right) = N_{\mathbb{K}}(\alpha_2 - \alpha_3).$$

Hence it can be expressed as $\lambda\eta$ for some unit η in the field and some fixed effectively computable λ. This gives

$$\lambda_1 \eta_1 + \lambda_2 \eta_2 + \lambda_3 \eta_3 = 0.$$

Thus by Theorem 3.1 there are only finitely many solutions η_1/η_3 and η_2/η_3; it follows that there are only finitely many x and y and these can be effectively determined. □

Theorem 3.3 holds similarly for the superelliptic equation $y^m = f(x)$ where $m \geq 3$ is a fixed integer and f has at least two simple zeros. In fact, on adopting the notation above, we immediately derive an equation of Thue type namely

$$\gamma_1 \zeta_1^m - \gamma_2 \zeta_2^m = \alpha_2 - \alpha_1$$

with coefficients and variables in $\mathbb{K}$, and it is easy to extend Theorem 3.2 to number fields; see [19]. Subsequent to the latter work, the hyperelliptic and superelliptic equations have been studied by many authors and they have given a variety of quantitative improvements and generalisations. We mention especially results of Brindza in this context, making effective a theorem of LeVeque of 1964 and extending the subject to S-integral solutions [56, 58].

Baker and Coates [29] proved in 1970 by way of Theorem 3.3 that there are only finitely many integer points on any algebraic curve of genus one and that these can, in principle, be effectively determined. They showed in fact that there exists a birational transformation taking the equation for the curve into hyperelliptic form and that the transformation can be chosen so as to be integral with respect to a number field. Their argument furnished an explicit upper bound for the sizes of the integer points on the curve and this was later improved by Kotov, Trelina, Schmidt and Bilu; see [123] for references. It remains an open problem to deal with arbitrary curves of higher genus. A famous theorem of Siegel [228] asserts that any algebraic curve of genus ≥ 1 has only finitely many integer points but the proof depends on the Mordell–Weil theorem to the effect that the group of rational points on the curve has a finite basis, as well as on the Thue–Siegel theory and, for reasons attaching to both of these topics, the result is not effective. In 1983, Faltings [90] obtained the celebrated theorem, originally conjectured by Mordell, that there are only finitely many rational points on any algebraic curve if the genus is ≥ 2. The proof is again non-effective, even in the case of integer points, but an essential component of Faltings' work has been made effective by Masser and Wüstholz [170] using the theory of logarithmic forms over algebraic groups; see the discussion in Chapter 7 and see also Vojta [250] for another approach based on Dyson [81].

For a wide ranging discussion on applications of our theory to discriminant form, index form and decomposable form equations of general type involving several variables, see again Győry [123]. To quote just two particular examples arising from this work, Győry has proved that an algebraic number field has only finitely many power integral bases $1, \alpha, \ldots, \alpha^{d-1}$ up to translation by elements of $\mathbb{Z}$ and, similarly up to translation, that there are only finitely many algebraic integers in the field with a given discriminant. Further, we remark that, as an early application of the p-adic theory of logarithmic forms, it was shown that one can determine, in principle, all elliptic curves with a given conductor (see [2] for an explicit computation in the case of conductor 11).

3.5 Practical computations

The method described in the preceding sections for the theoretical solution of Diophantine equations involves large numerical constants and some further work is necessary in order to complete the solution in particular cases.

The first result in this context is due to Baker and Davenport [30]. It was proved that the simultaneous Pell equations

$$3x^2 - 2 = y^2, \quad 8x^2 - 7 = z^2$$

have no solution in integers x, y, z except

$$(\pm 1, \pm 1, \pm 1), \quad (\pm 11, \pm 19, \pm 31).$$

Plainly a pair of simultaneous equations as above is essentially equivalent to a hyperelliptic equation

$$Y^2 = f(X)$$

with $X = x$, $Y = yz$ and so the method of Theorem 3.3 certainly applies in principle. The problem was to make the computations practical. The motivation for the study was a problem dating back to Diophantus based on the fact that the product of any two of 1, 3, 8, 120, increased by 1, is a perfect square; van Lint, recalling this problem in a lecture at Oberwolfach in 1968, asked whether the sequence can be continued to a fifth number and the solution to the equations above shows at once that such a number cannot exist. The result has generated a subject in its own right, namely the theory of Diophantine m-tuples; see Dujella [79, 80] and the references therein.

Subsequently Ellison *et al.* [84] used the same technique to show that the Mordell equation

$$y^2 = x^3 - 28,$$

which was then the smallest unsolved example, has only the solutions with (x, y) given by

$$(4, \pm 6), \quad (8, \pm 22), \quad (37, \pm 225).$$

The method has now become standard; we sketch a proof.

By factorisation over $\mathbb{Q}(\sqrt{-7})$ the equation

$$y^2 = x^3 - 28$$

becomes

$$\left(y + 2\sqrt{-7}\right)\left(y - 2\sqrt{-7}\right) = x^3$$

and so $y + 2\sqrt{-7} = \pm\zeta(u + v\xi)^3$, where u, v are integers, $\xi = \frac{1}{2}(1 + \sqrt{-7})$ and $\zeta = 1, 4/\xi$ or 2ξ. When $\zeta = 1$ we easily obtain $(u, v) = \pm(1, 4), \pm(5, -4)$ and these give $x = 37$. When $\zeta = 4/\xi$, on comparing coefficients of $\sqrt{-7}$, we get

$$u^3 - 6uv^2 - 2v^3 = \pm 2.$$

Moreover the case $\zeta = 2\xi$ also reduces to this equation. Thus on writing

$$u = 2Y, \quad v = -X - 2Y,$$

we see that it suffices to show that the Thue equation

$$X^3 - 12XY^2 - 12Y^3 = \pm 1$$

has only the solutions $(X, Y) = \pm(1, 0)$ and $\pm(1, -1)$; these give $x = 4$ and $x = 8$.

Now we work over the field $\mathbb{Q}(\vartheta)$ where $\vartheta = 2.7\cdots$ is a root of

$$\vartheta^3 - 12\vartheta - 12 = 0.$$

Fundamental units are given by $\eta_1 = -7 - 4\vartheta + \vartheta^2$ and $\eta_2 = 11 + \vartheta - \vartheta^2$ and we have

$$X - \vartheta Y = \pm \eta_1^{b_1} \eta_2^{b_2},$$

where b_1, b_2 are rational integers. We write $B = \max(|b_1|, |b_2|)$ and we assume that $B \geq 20$. Then, as in Theorem 3.2, we derive from the fact that the expression

$$(\vartheta^{(3)} - \vartheta^{(2)})(X - \vartheta^{(1)}Y) + (\vartheta^{(1)} - \vartheta^{(3)})(X - \vartheta^{(2)}Y)$$
$$+(\vartheta^{(2)} - \vartheta^{(1)})(X - \vartheta^{(3)}Y),$$

where the superscripts denote conjugates, vanishes identically, that we have

$$|b_1 \log \alpha_1 + b_2 \log \alpha_2 - \log \alpha_3| < e^{-(0.404)B},$$

where

$$\alpha_1 = \left| \frac{\eta_1^{(j)}}{\eta_1^{(k)}} \right|, \quad \alpha_2 = \left| \frac{\eta_2^{(j)}}{\eta_2^{(k)}} \right|, \quad \alpha_3 = \left| \frac{\vartheta^{(l)} - \vartheta^{(j)}}{\vartheta^{(k)} - \vartheta^{(l)}} \right|$$

and there are six possibilities for (j, k, l) with distinct elements.

The theory of linear forms in logarithms now gives $B < M$ where M was 10^{563} originally and is now (see Theorem 7.1) about 10^{25}. To complete the proof we set

$$\varphi = \frac{\log \alpha_1}{\log \alpha_2}, \quad \psi = \frac{\log \alpha_3}{\log \alpha_2}$$

and $\delta = 0.404$ so that

$$|b_1\varphi + b_2 - \psi| < e^{-\delta|b_1|}. \tag{3.1}$$

We select $K \geq 6$, for instance $K = 10^{33}$, and we determine from the continued fraction for φ integers p, q such that

$$|q\varphi - p| < 2/(KM), \quad 1 \leq q \leq KM.$$

We then invoke the following simple lemma.

Lemma 3.4 (Baker and Davenport) *If the distance of $q\psi$ from the nearest integer is at least $3/K$ then there is no solution of (3.1) in the range*

$$\delta^{-1} \log(K^2 M) < |b_1| < M.$$

Proof. If $|b_1|$ is in the above range then

$$e^{-\delta|b_1|} < 1/(K^2M);$$

hence from (3.1)

$$q|b_1\varphi + b_2 - \psi| < 1/K.$$

Further, we have

$$|b_1(q\varphi - p)| < 2/K.$$

Combining the inequalities we get

$$|pb_1 + qb_2 - q\psi| < 3/K$$

which contradicts the hypothesis. □

There is a very high probability, since there is no obvious connection between q and ψ, that the hypothesis of the lemma will be satisfied. On checking that it indeed holds for all six possible triples (j, k, l), one reduces the bound logarithmically, from 10^{563} to 3538 originally. A further application reduces it again to 44 and there is then no problem in determining all relevant b_1, b_2 and the corresponding solutions X, Y; the calculations originally involved estimating quantities to some 1000 decimal places but one could now work with around 50.

Many Diophantine equations have now been completely solved using this method; see again the survey article by Győry [123]. As he mentions, Bilu, Ellison, Gaál, Gebel, Hanrot, Herrmann, Heuberger, Lettl, Mignotte, Pethő, Pohst, Smart, Stroeker, Thomas, Tichy, Tzanakis, Voutier, Wakabayashi, de Weger, Wildanger and Zimmer, amongst others, have contributed significant results to this area. The works have involved two further computational techniques. Grinstead [120] derived a method involving recurrence sequences which was applied by Brown [59] and by Pethő [191] and further developed by Pinch [195]. Another development, due to de Weger [255] (see also Tzanakis and de Weger [246]), utilises an algorithm, the 'LLL-algorithm', due to A. K. Lenstra, H. W. Lenstra and L. Lovász [144]; this has improved considerably the computational aspects (see the survey article by Tijdeman [245] and also the book by Smart [232]). We give here just one example from this very large body of work, namely the remarkable application of Bilu,

Hanrot and Voutier [43] establishing a hundred year old conjecture that, for $n > 30$, the nth term of any Lucas or Lehmer sequence has a primitive divisor.

Recently there has been an exciting new development relating to elliptic curves based on the theory of linear forms in elliptic logarithms. It involves the group structure of the rational points on the curve and an analogue of Theorem 7.1 due to N. Hirata-Kohno and S. David. The method was indicated originally by Lang and first successfully applied by Stroeker and Tzanakis and, independently, by Gebel, Pethő and Zimmer; there is a discussion in Section 7.3. This area of research is now very active; a particularly striking result derived from the method, due to Gebel, Pethő and Zimmer [103], is a complete list of solutions of the Mordell equation $y^2 = x^3 + k$ for every k with $0 < |k| < 10^4$. Note the considerable advance on the small values of k that could be dealt with on an ad hoc basis before the advent of transcendence methods.

3.6 Exponential equations

It is a remarkable fact that many of the equations that we have considered above with fixed integer exponents can now be treated with variable exponents. The results arise from Sharpening I [23] where estimates for Λ that are best possible with respect to A were first obtained. All the work here is effective and, in contrast to Theorems 3.2, 3.3 and their generalisations, there is no analogous ineffective theory.

We begin by considering the equation

$$ax^n - by^n = c,$$

where a, b, c are given positive integers. When n is a given integer >2 the equation is of Thue type and we proved in Theorem 3.2 that there are only finitely many solutions in integers x and y all of which can be effectively bounded in terms of a, b, c and n. Now we can prove the following.

Theorem 3.5 *The equation above has only finitely many solutions in integers $x > 1$, $y > 1$ and $n > 2$ and they can in principle be effectively determined.*

Proof. We assume, as we may without loss of generality, that $y \geq x$. The equation in question can be written in the form $e^{\Lambda} - 1 = c/(by^n)$ where

$$\Lambda = \log(a/b) + n \log(x/y).$$

Since elementarily we have $|\log(1 + z)| \leq 2|z|$ if $|z| \leq \frac{1}{2}$, it follows on taking $z = e^{\Lambda} - 1$ that

$$|\Lambda| \ll y^{-n},$$

where, as later, the implied constant depends only on a, b and c. From Theorem 2.15 with $A_1 = \max(a, b)$, $A_2 = y$ and $B = n$ we have

$$\log |\Lambda| \gg -\log y \log n.$$

On cancelling $\log y$, we get $n \ll \log n$ and so n is bounded. It follows from Theorem 3.2 that also x and y are bounded and this establishes the result. □

Theorem 3.5 is the simplest deduction of its kind. Two deeper examples are given by the following.

Theorem 3.6 *The superelliptic equation $y^m = f(x)$ has only finitely many solutions in integers x, $y > 1$ and $m > 2$ and they can in principle be effectively determined.*

Here $f(x)$ denotes a polynomial with integer coefficients and with at least two simple zeros. The result is due to Schinzel and Tijdeman [206]. For the proof one observes, as in Section 3.4, that the superelliptic equation reduces to an equation of the type discussed in Theorem 3.5 with coefficients and variables in $\mathbb{K}$ and it is straightforward to generalise the argument to number fields.

Theorem 3.7 *For any integer $k \geq 6$, the equation*

$$y^m = 1^k + 2^k + \cdots + x^k + f(x)$$

has only finitely many solutions in integers $x \geq 1$, $y > 1$ and $m > 1$ and they can in principle be effectively determined.

In this case $f(x)$ is any polynomial with integer coefficients. The proof depends on properties of Bernoulli polynomials which enable one to reduce the equation to superelliptic type similar to that of Theorem 3.6. This was shown by Győry, Tijdeman and Voorhoeve [124] in 1979 but their work proceeded to utilise a theorem of LeVeque which had then only been proved ineffectively; an effective version was given by Brindza and this completed the proof of Theorem 3.7 (see [56, 57]). It is of interest to note that, if $k < 6$, then the equation of Theorem 3.7 can have infinitely many solutions. Indeed Schäffer [205] proved in 1956 that, when $f(x) = 0$ and $k > 0$, the equation has infinitely many solutions if and only if (k, m) is $(1, 2)$, $(3, 2)$, $(3, 4)$ or $(5, 2)$.

In 1976 Tijdeman [244] achieved a big success in this field by proving that the famous conjecture of Catalan is, in principle, decidable. Catalan conjectured in 1844 that the only solution in integers x, y, p, q, all >1, of the equation

$$x^p - y^q = 1$$

is given by $3^2 - 2^3 = 1$; thus he hypothesised that 8 and 9 are the only consecutive integer powers. In the Middle Ages, Levi ben Gerson solved the case $x = 3$, $y = 2$, in 1738 Euler solved the case $p = 2$, $q = 3$, in 1850 V. A. Lebesgue dealt with the equation $x^p - y^2 = 1$, in 1964 Chao Ko treated the more difficult example $x^2 - y^q = 1$ and in 1921 Nagell dealt with $x^3 - y^q = 1$ and $x^p - y^3 = 1$. Moreover Cassels showed in 1961 that if p, q are odd primes then p divides y and q divides x. Mąkowski deduced from the latter result that there cannot exist three consecutive integer powers. Ribenboim's book [201] gives references to all these works and covers the subject well. Tijdeman proved the following theorem.

Theorem 3.8 *There are only finitely many solutions of the Catalan equation $x^p - y^q = 1$ in integers x, y, p, q, all >1, and these can be effectively bounded.*

The original proof depended on a double application of a logarithmic form estimate of the type first obtained in Sharpening I (see Section 2.8) and the fact that it gave the best possible dependence on the maximum of the heights of the α was critical. Here we shall give a short demonstration using Theorem 2.15.

Proof of Theorem 3.8. Let x, y, p, q, all >1, be a solution of the Catalan equation. In view of the historical results mentioned above, it will suffice to assume that p, q are odd primes with $p > q$ and that $x > 2$. Then, by elementary factorisation, we have

$$x = kX^q + 1, \quad y = lY^p - 1$$

for some integers $X > 1$, $Y > 1$ where k is 1 or $1/p$ and l is 1 or $1/q$. It is readily verified that the equation $x^p - y^q = 1$ implies that $Y \ll X \ll p^{1/q}Y$ where, as later, the implied constants are absolute and we shall assume these estimates in the subsequent discussion.

Now the Catalan equation can be written in the form $x^p/y^q - 1 = 1/y^q$ and so, as in the proof of Theorem 3.5, we see that

$$|p \log x - q \log y| \ll y^{-q}.$$

We now substitute for x and y in terms of X and Y. Since

$$\left|\log x - \log kX^q\right| \ll (kX^q)^{-1}$$

and similarly for y and, further, both $q(lY^p)^{-1}$ and y^{-q} are at most $q^2Y^{-q} \ll p^3X^{-q}$, we get

$$|\Lambda| \ll p^3 X^{-q},$$

where

$$\Lambda = p \log k - q \log l + pq \log(X/Y).$$

We apply Theorem 2.15 with $n = 3$, $A_1 = p$, $A_2 = q$, $A_3 = \max(X, Y)$ and $B = pq$; then $A_3 \ll X$. This gives $\log|\Lambda| \gg -(\log p)^3 \log X$ and on using the upper bound for $|\Lambda|$ and cancelling $\log X$ it follows that $q \ll (\log p)^3$. Similarly on substituting for y only and leaving x unchanged we get

$$\left|\Lambda'\right| \ll q^2 Y^{-p},$$

where $\Lambda' = p\log(x/Y^q) - q\log l$. We now apply Theorem 2.15 again with $n = 2, A_1 = \max(x, Y^q), A_2 = q$ and $B = p$. Since $x \ll y^{q/p} \leq Y^q$, we have $A_1 \ll Y^q$ whence $\log\left|\Lambda'\right| \gg -q(\log p)^2 \log Y$ and on using the upper bound for $\left|\Lambda'\right|$ and cancelling $\log Y$ we obtain $p \ll q(\log p)^2$.

Combining the latter inequality with $q \ll (\log p)^3$ we see that p and q are bounded and it follows from Theorem 3.6 that also x and y are bounded. □

Since the work of Tijdeman there has been considerable effort to obtain manageable bounds for the solutions of the Catalan equation so as to verify Catalan's conjecture. The best estimates to date show that $\min(p, q)$ lies between 10^7, due to Mignotte *et al.*, and 10^{12}, due to Blass, Glass *et al.*, but it has proved difficult to close the gap. Recently, however, there has been some very exciting new work of Mihăilescu [178] that completely solves the problem. First, following on work of Inkeri and Mignotte, he showed [177] that p, q must satisfy a double Wieferich condition, that is

$$p^{q-1} \equiv 1 (\bmod\ q^2), \quad q^{p-1} \equiv 1 (\bmod\ p^2).$$

Using this result together with new ideas from the theory of cyclotomic fields, in particular results relating to cyclotomic units based on a deep theorem of Thaine, Mihăilescu has eliminated the case $p \not\equiv 1$ (mod q); there remains the case $p \equiv 1$ (mod q) and here Mihăilescu shows that the classical paper of Cassels and a memoir of Hyyrö of about the same time, combined with the fundamental results on logarithmic forms, can be applied to complete the solution. For a rendition of Mihăilescu's work which appeared in the Bourbaki seminar series before his paper [178] was published see Bilu [42] and for a significant later development see [179].

For further work in connection with exponential Diophantine equations we refer to the excellent tract by Shorey and Tijdeman [225]. They discuss there, in particular, perfect powers in binary recurrence sequences, generalised hyperelliptic and superelliptic equations relating to the p-adic domain and to algebraic number fields, and to various famous equations in addition to that of Catalan. These include the equation

$$\binom{x+n}{n} = y^m$$

which was solved by Erdős in 1951, the equation

$$(x+1)\cdots(x+n) = y^m$$

which was solved by Erdős and Selfridge in 1975 and the celebrated Fermat equation

$$x^n + y^n = z^n$$

which was later solved through a particular case of the Taniyama–Shimura conjecture on the modularity of elliptic curves by Wiles [256] in 1995.

3.7 The *abc*-conjecture

The *abc*-conjecture is a simple statement about integers that has come to be recognised as one of the key problems of mathematics. It has its origin in a conjecture of Szpiro comparing the discriminant and conductor of an elliptic curve; on taking the special case of the so-called Frey curve $y^2 = x(x+a)(x-b)$, which, incidentally, underlies Wiles' proof of Fermat's last theorem referred to above, Oesterlé formulated an assertion about the sizes of relatively prime non-zero integers a, b, c satisfying

$$a + b + c = 0,$$

and this was refined by Masser [163] to give the following.

Conjecture 3.9 (Oesterlé–Masser) *For any $\varepsilon > 0$ we have*

$$\max(|a|, |b|, |c|) \ll N^{1+\varepsilon},$$

where the implied constant depends only on ε.

Here N denotes the 'conductor' or 'radical' of abc, that is the product of all the distinct prime factors of abc. The conjecture would not hold with $\varepsilon = 0$ since for instance (see [139]), on taking $a = 1, b = -3^{2^n}$, and writing $3 = 1+2$, it is clear that 2^n divides c whence $N \leq 6c/2^n$. Masser was influenced by a theorem of Mason [161] who had investigated the problem of generalising the theory of logarithmic forms to function fields and was led thereby to a result which, as we now see, is the analogue of the *abc*-conjecture in the function field setting; Stothers [239] had independently come upon the same result by way of the theory of Riemann surfaces.

The conjecture has many striking consequences and has been discussed widely; see in particular [119]. One observes immediately that it implies that the Fermat–Catalan equation

$$ax^r + by^s + cz^t = 0,$$

where a, b, c are given non-zero integers, has only finitely many solutions in relatively prime integers x, y, z, all > 1, and positive exponents r, s, t satisfying $\lambda < 1$, where $\lambda = (1/r) + (1/s) + (1/t)$. In fact Conjecture 3.9 gives $\max(x^r, y^s, z^t) \ll (xyz)^{1+\varepsilon}$ where the implied constant depends on a, b, c and ε; thus we obtain $x \ll (xyz)^{(1+\varepsilon)/r}$ and similarly for y, z whence $xyz \ll (xyz)^{(1+\varepsilon)\lambda}$. By an observation of Masser, for $\lambda < 1$ the largest value of λ as a function of positive integers r, s and t is attained at $r = 2$, $s = 3$, $t = 7$ and then $\lambda = 41/42$; thus with $\varepsilon < (1/41) \leq (1/\lambda) - 1$ it follows that xyz, whence also each of x, y, z, is bounded in terms of a, b, c and the estimates above now show that r, s, t are likewise bounded. In other directions, Elkies has demonstrated that the *abc*-conjecture furnishes a proof of the famous theorem of Faltings (see Section 3.4) and, furthermore, Langevin and Bombieri have shown that it can be used to establish the Thue–Siegel–Roth theorem (see Section 1.1 and [48, Theorem 12.2.9]). The latter arguments utilise a theorem of G. V. Belyĭ [37] on coverings of algebraic curves with three ramification points but are otherwise quite straightforward. To obtain Theorem 1.7 in the special case of fractional powers of rationals $(a/b)^{1/n}$, for example, one has only to apply Conjecture 3.9 to the sum $c = bx^n - ay^n$ to get $|c| \gg y^{n-2-\varepsilon}$. Granville [117] has demonstrated that the *abc*-conjecture yields various results on square-free numbers of a kind studied by Hooley and others. Moreover Granville and Stark [118] have shown that, if one assumes a so-called uniform *abc*-conjecture for number fields as formulated in studies of Vojta [249], then one can prove the non-existence of the Siegel zero for Dirichlet L-functions (see Section 3.1); and there is a connection here, through work of Baker and Schinzel on the genera of binary quadratic forms, with the famous 'numeri idonei' problem of Euler (see the survey [34]).

The only significant approach to date to the *abc*-conjecture is due to Stewart and Kunrui Yu [238], refining earlier work of Stewart and

Tijdeman [237]. They show that

$$\log \max(|a|, |b|, |c|) \ll N^{1/3}(\log N)^3,$$

where the implied constant is absolute. The result is based on the archimedean estimate for logarithmic forms in Theorem 7.1 and the non-archimedean analogues of Kunrui Yu [268] as discussed in Section 2.8. The approach is plainly of much interest but it would seem that, to go significantly further, some new ideas will be needed and certain suggestions in this connection were made recently in Baker [27, 28]. Let $\omega(n)$ denote the number of distinct prime factors of an integer n and define $\omega = \omega(abc)$. Then it was proposed that, for all relatively prime integers a, b, c with $a + b + c = 0$, we have

$$\max(|a|, |b|, |c|) \ll N(\log N)^{\omega}/\omega!,$$

where the implied constant is absolute. Since $\omega \ll \log N/\log\log N$, it is seen that this is a refinement of Conjecture 3.9. Moreover, it does not violate the example with $a = 1$, $b = -3^{2^n}$ mentioned at the beginning or indeed a stronger result of Stewart and Tijdeman [237] to the effect that there exist infinitely many positive a, b, c, with no common factor and $a - b = c$, such that

$$\log(c/N) \gg (\log N)^{1/2}/\log\log N.$$

Furthermore, computations as described in [28] indicate that the refinement may be valid in a completely explicit form with 6/5 as the implied constant. There is another refinement to the *abc*-conjecture proposed in [27, 28] namely

$$\max(|a|, |b|, |c|) \ll N\Theta(N),$$

where $\Theta(N)$ denotes the number of positive integers up to N that are composed only of prime factors of N, and again computations have been described providing support.

Now consider the logarithmic form

$$\Lambda = u_1 \log v_1 + \cdots + u_n \log v_n,$$

where $v_1, \ldots, v_n$ are positive integers and $u_1, \ldots, u_n$ are integers, not all 0. Assuming that $\Lambda \neq 0$ we can write $\Lambda = \log(a/b)$ for unique positive integers a, b with $(a,b) = 1$. We put $c = a - b$, and we note that, if p is a prime dividing c, then $\log(a/b)$ exists in the p-adic sense and we have $|\Lambda|_p = |c|_p$; if p does not divide c we simply define $|\Lambda|_p = 1$. Then in the case that $a = b + c$ with a, b, c positive, a slight variant of the proposed refinements above is equivalent to an estimate for the expression

$$\Xi = \min(1, |\Lambda|) \prod \min(1, p|\Lambda|_p),$$

with the product taken over all primes p; in a little weaker form, this estimate is given by

$$\log \Xi \gg -(\log v_1 + \cdots + \log v_n) \log u,$$

where $u = \max |u_j|$. Thus we see that a result of the strength of the *abc*-conjecture sufficient for all its main applications amounts essentially to (i) replacing the archimedean valuation $|\Lambda|$ in Theorem 7.1 (or Theorem 2.15) by Ξ, together with (ii) replacing the product $h'(\alpha_1) \cdots h'(\alpha_n)$ of the heights of the α by the sum $h'(\alpha_1) + \cdots + h'(\alpha_n)$ (or, in Theorem 2.15, the product of the $\log A_j$ by their sum). Certainly combining the archimedean and non-archimedean valuations in a way suggested by the product formula in algebraic number theory would be of much value.

The subsequent chapters are devoted to the theory of logarithmic forms in the context of elliptic and abelian functions and more generally algebraic groups. Before leaving the subject of Diophantine problems and the classical theory, however, we should mention that the work has thrown valuable light on many further topics. They include Bertrand's studies on Galois representations [38], Bilu's contributions to the theory of modular curves [41] (see also Odoni [189] on modular forms), results of Murty and others on the Ramanujan τ-function [185] and they have found application to dynamical systems (see K. Schmidt [207]) and even to knot theory (see Riley [202], Acuña and Short [1]).

4
Commutative algebraic groups

4.1 Introduction

This chapter will be devoted to the preliminary results on commutative group varieties needed for the proof of the fundamental multiplicity estimates that we shall give in Chapter 5. The estimates are crucial to the recent refined results concerning linear forms in logarithms; essentially, they replace the generalised van der Monde determinants and the Kummer descent as used in the original work. The theorems in Chapter 5 will furnish all that is needed for the treatment of logarithmic forms in both the homogeneous and inhomogeneous cases as well as dependence relations of the type discussed in Section 7.1. The approach given here employs techniques from commutative algebra and algebraic geometry and extends to much more general situations, in particular to arbitrary algebraic groups. The basic tools used in the derivation of the multiplicity estimates are the theory of Hilbert functions and the intersection theory of varieties (see Wüstholz [263]).

Before going into details we give a short historical account of the use of these methods in transcendence. They were in fact introduced by Nesterenko [186] in 1977 in connection with his studies on E-functions and then further developed by Brownawell and Masser [61]. A variant of this technique combining commutative algebra with certain arithmetical considerations was given by Wüstholz [258] in 1980. Subsequently, between 1981 and 1985, Masser and Wüstholz [164, 165, 166] proved a zero estimate on group varieties appertaining to a single differential operator. The main problem then became to extend the result to arbitrarily many operators and Wüstholz succeeded in solving the problem in

1982; the work was presented in his Habilitationsschrift *Neue Methoden in der Theorie der transzendenten Zahlen* of the University of Wuppertal. The first published paper in this context, however, is due to Philippon [193]; in an important work of 1986 he was able to give a result of the same kind but with elimination of certain restrictions on the parameters and indeed in a form that is close to best possible. Later, in [262] and [263], Wüstholz himself published detailed and improved versions of his original material. The theory has found numerous applications as we shall describe in subsequent chapters.

Many results in this field can be expressed in terms of exponential maps on certain commutative algebraic groups; an observation of this kind was indeed noted by Lang [134, 137] in the early 1960s. The terminology leads to a unification of several classical transcendence proofs including those of Hermite, Lindemann, Gelfond and Schneider. Essentially two properties of the fundamental functions are used, namely the differential equations and the addition laws that the functions satisfy. Thus, for instance, Schneider's solution to Hilbert's seventh problem can be described as the construction of a polynomial $P(X, Y)$ which vanishes at points of the form $(k + l\beta, \alpha^{k+l\beta})$ for a certain set of integers k, l under the assumption that α^{β} is algebraic. The range of the set is then enlarged by an extrapolation process based on the maximum-modulus principle and it is then shown that there cannot be too many zeros of the above type; for the last step Schneider simply appeals to the non-vanishing of a van der Monde determinant. By contrast, in the approach by zero estimates, a polynomial $P(X, Y)$ is constructed such that, for integers k, l as above, the polynomials

$$P_{k,l}(X, Y) = P(X + k + l\beta,\ \alpha^{k+l\beta} Y),$$

vanish at $(X, Y) = (0, 1)$; this is obviously just another way of viewing the earlier construction. By a so-called Schwarz lemma, itself a consequence of the maximum-modulus principle, it is deduced that $P_{k,l}(0, 1) = 0$ for an extended range of integers k, l, that is, one obtains a large set of apparently distinct polynomials with a common zero. A suitable zero estimate now shows that this is impossible; in fact, provided that P does not vanish identically, one sees that the polynomials generate an ideal in $\mathbb{C}[X, Y]$ which cannot have a zero as soon as k, l vary in a sufficiently large set.

Similarly, in Gelfond's solution to Hilbert's seventh problem, a polynomial $P(X, Y)$ is constructed such that the function $f(z) = P(e^z, e^{\beta z})$ does not vanish identically and has a zero at $z = s\log\alpha$ with high multiplicity, that is, one has $f^{(t)}(s\log\alpha) = 0$ for a large set of integers s, t. The range for t is then extended by an interpolation process and finally Gelfond appeals to the non-vanishing of a van der Monde determinant as in the Schneider proof (see Chapter 2). Here one can use instead a multiplicity estimate in the same way as the zero estimate previously. Accordingly, one constructs a polynomial $P(X, Y)$ such that, for integers s, t in a certain range, the derived polynomials

$$P_{s,t}(X, Y) = (\Delta^t P)(\alpha^s X, \alpha^{s\beta} Y)$$

vanish at $(X, Y) = (1, 1)$, where $\Delta = X\frac{\partial}{\partial X} + \beta Y\frac{\partial}{\partial Y}$. The domain is expanded by interpolation and, by considering the ideal generated by the $P_{s,t}$ in $\mathbb{C}[X, Y]$, a contradiction is obtained as before.

More generally, Baker's theorem can be based on the multiplicity estimates proved in Chapter 5. Following Wüstholz [262], a polynomial $P(X_0, X_1, \ldots, X_n)$ with algebraic coefficients is constructed such that, for integers s, $t_0, \ldots, t_{n-1}$ in a certain domain, the derived polynomials

$$\begin{aligned} P_{s,t_0,t_1,\ldots,t_{n-1}}&(X_0, X_1, \ldots, X_n) \\ &= (\Delta_0^{t_0}\cdots\Delta_{n-1}^{t_{n-1}}P)\,(X_0 + s,\ \alpha_1^s X_1, \ldots, \alpha_n^s X_n) \end{aligned}$$

vanish at the point $(0, 1, \ldots, 1)$; here $\alpha_1, \ldots, \alpha_n$, β_0, $\beta_1, \ldots, \beta_{n-1}$ are algebraic numbers and

$$\Delta_0 = \frac{\partial}{\partial X_0} + \beta_0 X_n\frac{\partial}{\partial X_n}, \quad \Delta_j = X_j\frac{\partial}{\partial X_j} + \beta_j X_n\frac{\partial}{\partial X_n} \quad (1 \le j < n).$$

By an extrapolation algorithm as described in Chapter 2 one deduces that the derived polynomials taken over a larger domain again vanish at $(0, 1, \ldots, 1)$. Now the domain is sufficiently large relative to the degree of P so that the multiplicity estimates imply that some degeneracies occur. The latter arise in a natural way as algebraic subgroups of the algebraic group $G = \mathbb{G}_a \times \mathbb{G}_m^n$, where $\mathbb{G}_a$ and $\mathbb{G}_m$ are the additive and multiplicative group varieties. They can be described easily by means of the Geometry of Numbers and correspond to a necessary hypothesis on the linear independence of either the $\log \alpha$ or the β as one finds in Baker's theorem.

4.2 Basic concepts in algebraic geometry

We summarise here some of the rudiments of algebraic geometry; for a full exposition see, for example, Hartshorne [125, Ch. I]. We define the affine n-space $\mathbb{A}^n$ over an algebraically closed field $\mathbb{K}$ of characteristic 0 as the set of all n-tuples of elements of $\mathbb{K}$. By an algebraic set over $\mathbb{K}$ we mean a subset of $\mathbb{A}^n$ given by the zeros of a collection of elements in the ring $\mathbb{K}[X_1, \ldots, X_n]$ of polynomials in n variables; in our subsequent discussion all the algebraic sets will be assumed as defined over $\mathbb{K}$ and we shall omit reference to the field. An algebraic set is called irreducible if it is not empty and cannot be decomposed into the union of two proper algebraic sets; in this sense, the empty set is not considered to be irreducible. The algebraic sets generate a topology, the Zariski topology, such that the open sets are given by the complements of the algebraic sets.

An affine variety X over $\mathbb{K}$, or briefly affine variety, is an irreducible algebraic subset of an affine space $\mathbb{A}^n$ for some integer n. From the irreducibility of X it follows that an affine variety is given by the set of zeros of some collection of polynomials in $\mathbb{K}[X_1, \ldots, X_n]$, with the property that the ideal $I(X)$ generated by the collection is a prime ideal. The affine coordinate ring $A[X]$ of X is defined as the quotient ring $\mathbb{K}[X_1, \ldots, X_n]/I(X)$.

Let now Y be an open subset of an affine variety in $\mathbb{A}^n$; it is customary to refer to such a subset Y as a quasi-affine variety. From the general theory in [125] one sees that any quasi-affine variety is in fact isomorphic to an affine variety but we do not need to appeal to that result here. A function $f : Y \to \mathbb{K}$ is said to be regular at a point in Y if there is an open neighbourhood U of the point such that f can be written there as a quotient p/q with p, q polynomials in $\mathbb{K}[X_1, \ldots, X_n]$ and with q nowhere zero on U. If f is regular at every point in Y then it is termed a regular function on Y. The regular functions on Y form a ring $\mathcal{O}(Y)$, indeed an algebra, and it is a basic theorem of algebraic geometry (again see [125]) that, for an affine variety X, we have $\mathcal{O}(X) = A[X]$, the coordinate ring of X.

A morphism $\varphi: X \to Y$ between affine or quasi-affine varieties X, Y is defined as a continuous map such that, for every regular function $f: V \to \mathbb{K}$, where V is an arbitrary open subset in Y, the composition $\varphi^*(f)$ of f and φ (i.e. $f \circ \varphi$ mapping $\varphi^{-1}(V)$ to $\mathbb{K}$) is regular. Clearly $\varphi^*(f)$ is in $\mathcal{O}(X)$ for f in $\mathcal{O}(Y)$ and hence any morphism φ induces a ring homomorphism φ^* between $\mathcal{O}(X)$ and $\mathcal{O}(Y)$; this is termed the comorphism of φ. We shall be concerned in the sequel with coordinate rings and accordingly we mention that defining an affine variety X over $\mathbb{K}$ amounts to defining a finitely generated integral domain over $\mathbb{K}$ and, moreover, if X, Y are affine varieties, then specifying a morphism φ from X to Y amounts to specifying a comorphism φ^* from $A[Y] = \mathcal{O}(Y)$ to $A[X] = \mathcal{O}(X)$; these facts follow from [125, Corollary I.3.8].

4.3 The groups $\mathbb{G}_a$ and $\mathbb{G}_m$

The additive group variety $\mathbb{G}_a$ is defined as the affine line $\mathbb{A}^1$ together with the group laws $\mu(g,h) = g + h$ and $\iota(g) = -g$. From Section 4.2, we see that $\mathcal{O}(\mathbb{G}_a)$ is the polynomial ring $\mathbb{K}[X]$. Further, the comorphism μ^* corresponding to μ maps $\mathbb{K}[X]$ to $\mathbb{K}[X_1, X_2]$ by $\mu^*(X) = X_1 + X_2$, where X_1 and X_2 are the elements of $\mathcal{O}(\mathbb{G}_a \times \mathbb{G}_a)$ which take (g,h) into g and h respectively. Similarly the comorphism corresponding to ι, taking $\mathbb{K}[X]$ to $\mathbb{K}[X]$, is given by $X \mapsto -X$. It is readily seen that, with these definitions, $\mathbb{G}_a$ becomes a commutative group variety.

Since $\mathbb{G}_a$ as a variety is just the affine line $\mathbb{A}^1$, it can be embedded into the projective space $\mathbb{P}^1$. A particularly nice embedding can be described through the right group operation of PGL_2 on $\mathbb{P}^1$. That is, we can identify $\mathbb{G}_a$ with a subgroup of PGL_2 by the mapping

$$t \longmapsto \begin{pmatrix} 1 & t \\ 0 & 1 \end{pmatrix}$$

and then the embedding $\mathbb{G}_a \to \mathbb{P}^1$ is given by $t \mapsto (1, t+1)$, where the latter is the image of $(1, 1)$ under the aforementioned operation,

$$(1, t+1) = (1, 1) \begin{pmatrix} 1 & t \\ 0 & 1 \end{pmatrix}.$$

In this way $\mathbb{G}_a$ becomes the orbit of $(1, 1)$ in $\mathbb{P}^1$ via the right action of the subgroup of PGL_2 on $\mathbb{P}^1$.

Analogously, the multiplicative group variety $\mathbb{G}_m$ is defined as the affine line with the origin removed, that is $\mathbb{A}^1 \smallsetminus \{0\}$, together with the group laws $\mu(g,h) = gh$ and $\iota(g) = g^{-1}$; thus $\mathbb{G}_m$ is quasi-affine with respect to $\mathbb{A}^1$. Further, we have $\mathcal{O}(\mathbb{G}_m) = \mathbb{K}[X, X^{-1}]$; for the non-trivial closed sets in the Zariski topology of $\mathbb{A}^1$ have just finitely many points, whence any element f in $\mathcal{O}(\mathbb{G}_m)$ has the form $p(X)/q(X)$ with p, q polynomials and q nowhere zero on $\mathbb{G}_m$ and the latter implies that q is monomial. (Alternatively, $\mathbb{G}_m$ can be defined as the hyperbola with equation $ST = 1$ in the affine plane $\mathbb{A}^2$; for this is plainly isomorphic to $\mathbb{A}^1 \smallsetminus \{0\}$ by projection onto the affine X-line. Now the ideal $I(\mathbb{G}_m)$ is generated by $XY - 1$ and it is clear that the quotient ring $\mathbb{K}[X, Y]/(XY - 1)$ is isomorphic to the ring $\mathbb{K}[X, X^{-1}]$.)

Next we observe that the comorphism μ^* corresponding to μ maps $\mathbb{K}[X, X^{-1}]$ to $\mathbb{K}[X_1, X_1^{-1}, X_2, X_2^{-1}]$ by $\mu^*(X) = X_1X_2$, where X_1, X_2 are the elements of $\mathcal{O}(\mathbb{G}_m \times \mathbb{G}_m)$ defined as for $\mathbb{G}_a$. Further, the comorphism corresponding to ι, taking $\mathbb{K}[X, X^{-1}]$ to $\mathbb{K}[X, X^{-1}]$, is given by $X \mapsto X^{-1}$. Obviously $\mathbb{G}_m$ can be embedded into $\mathbb{G}_a$ and, as before, this gives a projective embedding of $\mathbb{G}_m$ into $\mathbb{P}^1$. Another embedding $\mathbb{G}_m \to \mathbb{P}^1$ is given by $t \mapsto (1, t)$, where $\mathbb{G}_m$ is identified with a subgroup of PGL_2 by the mapping

$$t \longmapsto \begin{pmatrix} 1 & 0 \\ 0 & t \end{pmatrix}.$$

Since we have

$$(1, t) = (1, 1) \begin{pmatrix} 1 & 0 \\ 0 & t \end{pmatrix}$$

it follows that $\mathbb{G}_m$, like $\mathbb{G}_a$, is the orbit of $(1, 1)$ in $\mathbb{P}^1$.

Finally, on combining results, we see that for the group variety $G = \mathbb{G}_a \times \mathbb{G}_m^n$, the ring $\mathbb{K}[G]$ of regular functions on G is given by

$$\mathbb{K}[X_0, X_1, X_1^{-1}, \ldots, X_n, X_n^{-1}].$$

Further, the embeddings of $\mathbb{G}_a$ and $\mathbb{G}_m$ into $\mathbb{P}^1$ described above plainly give an embedding of G into $(\mathbb{P}^1)^{n+1}$.

4.4 The Lie algebra

In this section we determine the Lie algebra $\mathfrak{g} = \mathfrak{g}(G)$ of commutative linear algebraic groups G which are products of the groups $\mathbb{G}_a$ and $\mathbb{G}_m$ that were introduced in Section 4.3.

The morphism $\mu: G \times G \to G$ is the product of the morphisms which define the group structures on $\mathbb{G}_a$ and $\mathbb{G}_m$. We define the translation operator T_g for any $g \in G$ as the mapping $G \to G$ which takes h in G to $\mu(g, h)$. Then the comorphism T_g^* of T_g takes $\mathcal{O}(G)$ into itself and, for any element f in $\mathcal{O}(G)$, we have $T_g^*(f)$ as the function $G \to \mathbb{K}$ given by $h \mapsto f(\mu(g, h))$ for $h \in G$. We recall that, associated with $\mathcal{O}(G)$, there is the vector space of derivations over $\mathbb{K}$. An element $\mathcal{D}: \mathcal{O}(G) \to \mathcal{O}(G)$ of this vector space is called translation invariant if

$$\mathcal{D} \circ T_g^* = T_g^* \circ \mathcal{D}$$

for all $g \in \mathbb{G}$. The Lie algebra $\mathfrak{g} = \mathfrak{g}(G)$ of G is now defined to be the vector space of translation invariant derivations of $\mathcal{O}(G)$.

Accordingly the Lie algebras $\mathfrak{g}_a$ and $\mathfrak{g}_m$ of $\mathbb{G}_a$ and $\mathbb{G}_m$ are the spaces of translation invariant derivations of the algebras $\mathbb{K}[X]$ and $\mathbb{K}[X, X^{-1}]$ respectively. They can be determined easily in terms of the derivation d/dX which is defined by $(d/dX)X = 1$.

Proposition 4.1 *The algebras $\mathfrak{g}_a$ and $\mathfrak{g}_m$ are generated over $\mathbb{K}$ by d/dX and by $X(d/dX)$ respectively.*

Proof. First we note that a derivation of $\mathbb{K}[X]$ is completely determined by its action on X and similarly for $\mathbb{K}[X, X^{-1}]$. Further, clearly, for $\mathbb{G}_a$ we have $T_g^*(X) = g + X$ and for $\mathbb{G}_m$ we have $T_g^*(X) = gX$. Hence the action of d/dX on $T_g^*(X)$ gives 1 for $\mathbb{G}_a$ and g for $\mathbb{G}_m$.

We proceed now to prove the proposition for $\mathfrak{g}_a$. Accordingly let $\mathcal{D}$ be any element in $\mathfrak{g}_a$. Then $\mathcal{D}X \in \mathbb{K}[X]$ and we have

$$\mathcal{D}(T_g^*(X)) = \mathcal{D}(g + X) = \mathcal{D}X\,.$$

Since $\mathcal{D}$ is translation invariant, it follows that

$$T_g^*(\mathcal{D}X) = \mathcal{D}X$$

whence, for all g, the polynomial $\mathcal{D}X$ is invariant under the substitution $X \mapsto g + X$. This implies that $\mathcal{D}X \in \mathbb{K}$ and so $\mathcal{D}$ is in the vector space generated over $\mathbb{K}$ by d/dX as required.

Suppose now that $\mathcal{D}$ is an element of $\mathfrak{g}_m$. Then $\mathcal{D}X \in \mathbb{K}[X, X^{-1}]$ whence $X^n\mathcal{D}X$ is a polynomial, say $P(X)$, for some integer n, and we may select n such that $(X, P(X)) = 1$. We have

$$T_g^*(\mathcal{D}X) = (\mathcal{D}X)(gX), \quad \mathcal{D}(T_g^*(X)) = g\mathcal{D}X .$$

Since $\mathcal{D}$ is translation invariant, it follows that

$$(gX)^{-n}P(gX) = gX^{-n}P(X) .$$

But this gives $P(gX) = g^{n+1}P(X)$ which implies that P is homogeneous of degree $n+1$. Thus we obtain $P(X) = P(1)X^{n+1}$ and so, from the property $(X, P(X)) = 1$, we see that $n+1 = 0$. Hence $\mathcal{D}X = P(1)X$ and therefore $\mathcal{D}$ has the form $P(1)X\,(d/dX)$ as required. □

In a similar way one deduces that the space of translation invariant 1-forms is generated by dX in the case of $\mathbb{G}_a$ and by dX/X in the case of $\mathbb{G}_m$.

Finally we note that the Lie algebra of $G = \mathbb{G}_a \times \mathbb{G}_m^n$ is the direct sum of the Lie algebras of the factors. Thus it follows from the proposition that the differential operators

$$\frac{\partial}{\partial X_0}, \; X_1\frac{\partial}{\partial X_1}, \dots, X_n\frac{\partial}{\partial X_n}$$

form a basis for $\mathfrak{g}$ and that the differentials

$$dX_0, \; \frac{dX_1}{X_1}, \dots, \frac{dX_n}{X_n}$$

give a basis for its dual space $\mathfrak{g}^* = \mathrm{Hom}_{\mathbb{K}}(\mathfrak{g}, \mathbb{K})$, the space of translation invariant derivations. These bases for $\mathfrak{g}$ and $\mathfrak{g}^*$ are plainly dual and the choice of bases gives isomorphisms between $\mathfrak{g}$ and $\mathbb{K}^{n+1}$ and between $\mathfrak{g}^*$ and $\mathbb{K}^{n+1}$.

4.5 Characters

Let G, G' be group varieties of the type $\mathbb{G}_m^n$ or $\mathbb{G}_a \times \mathbb{G}_m^n$ for some $n \geq 0$. A homomorphism from G to G' is a morphism $\varphi\colon G \to G'$ such that the diagram

$$\begin{array}{ccc} G \times G & \xrightarrow{\varphi \times \varphi} & G' \times G' \\ {\scriptstyle m}\downarrow & & \downarrow{\scriptstyle m'} \\ G & \xrightarrow[\varphi]{} & G' \end{array}$$

commutes where m, m' are the multiplication morphisms for the groups G and G' respectively. Then the group of multiplicative characters of G is defined to be

$$X(G) = \mathrm{Hom}(G, \mathbb{G}_m).$$

Further

$$X_*(G) = \mathrm{Hom}(\mathbb{G}_m, G)$$

is called the group of multiplicative one-parameter subgroups of G. The character groups $X(\mathbb{G}_m)$ and $X(\mathbb{G}_a)$ are easily determined.

Proposition 4.2 *We have*

(i) $X(\mathbb{G}_m) \cong \mathbb{Z}$,
(ii) $X(\mathbb{G}_a) = 0$,
(iii) $X_*(\mathbb{G}_a) = 0$.

Proof. To prove (i) let $\lambda \in X(\mathbb{G}_m)$. Then λ^* is an endomorphism of $\mathbb{K}[T, T^{-1}]$. Such a homomorphism is uniquely determined by its effect on the generator T. So let

$$\lambda^*(T) = T^n q(T)$$

with $q(T)$ in $\mathbb{K}[T], n \in \mathbb{Z}$ and $(T, q(T)) = 1$. On noting that the affine algebra of $\mathbb{G}_m \times \mathbb{G}_a$ is the tensor product of the affine algebras of the factors, we have on the one hand

$$\begin{aligned}(\lambda^* \otimes \lambda^*)(m^*T) &= (\lambda^* \otimes \lambda^*)(T_1 \otimes T_2)\\ &= \lambda^*(T_1) \otimes \lambda^*(T_2)\\ &= T_1^n q(T_1) \otimes T_2^n q(T_2)\end{aligned}$$

and, on the other hand,

$$\begin{aligned}m^*\lambda^*(T) &= m^*\big(T^n q(T)\big)\\ &= m^*(T^n) \cdot m^*(q(T))\\ &= (T_1 \otimes T_2)^n \cdot q(T_1 \otimes T_2)\,.\end{aligned}$$

Since λ^* is a ring homomorphism both are equal and therefore

$$q(T_1 \otimes T_2) = q(T_1) \otimes q(T_2)\,.$$

This functional equation for $q(T)$ has the solutions

$$q(T) = T^m$$

for integers $m \geq 0$. But $(q, T) = 1$ and so $q(T) = 1$. Hence we see that $\lambda^*(T) = T^n$ for some $n \in \mathbb{Z}$.

In this way we obtain a map $\lambda \mapsto n = n(\lambda)$ from $X(\mathbb{G}_m)$ to $\mathbb{Z}$ which is clearly a homomorphism that is trivially injective and surjective. This proves (i). Note that it follows immediately that $X(\mathbb{G}_m^n) \cong \mathbb{Z}^n$.

In order to prove (ii) we take λ in $X(\mathbb{G}_a)$ and obtain an algebraic homomorphism $\lambda^*\colon \mathbb{K}[T, T^{-1}] \to \mathbb{K}[T]$. But $\lambda^*(T)$ has to be invertible which means that λ^* is a unit in $\mathbb{K}[T]$. This shows that λ is constant and being a homomorphism it must be zero. Part (iii) of the proposition can be seen as follows. If there were a non-trivial one-parameter subgroup then the additive group would contain torsion subgroups which is not the case. □

The character group can be used to give a more intrinsic description for the affine algebra of $G = \mathbb{G}_m^n$. Let $\mathfrak{g}$ be the Lie algebra of G as discussed in Section 4.4. We write $\mathcal{T}$ instead of G and then its affine algebra is the group algebra $\mathbb{K}[X(\mathcal{T})]$ of the character group. This description can be used to establish a canonical identification of the character group $X(\mathcal{T})$ with a lattice in $\mathfrak{g}^* = \mathrm{Hom}_{\mathbb{K}}(\mathfrak{g}, \mathbb{K})$, that is a a free subgroup of $\mathfrak{g}^*$ of maximal rank, and likewise of the co-character group $X_*(\mathcal{T})$ with a lattice in $\mathfrak{g}$. We associate with $\chi \in X(\mathcal{T})$ the invariant 1-form $l_\chi = d\chi/\chi$ and we associate with $\psi \in X_*(\mathcal{T})$ the invariant differential operator ∂_ψ defined by $(d\psi)(\chi \cdot (d/d\chi))$ where χ is one of the two generators for the character group $X(\mathbb{G}_m)$; thus we obtain maps $l: X(\mathcal{T}) \to \mathfrak{g}^*$, $l \mapsto l_\chi$, and $\partial: X_*(\mathcal{T}) \to \mathfrak{g}$, $\psi \mapsto \partial_\psi$. We observe here that ∂ is canonical up to the choice of χ or its inverse χ^{-1} and moreover that l and ∂ give an identification of $X(\mathcal{T})$ with a lattice in $\mathfrak{g}^*$ and of $X_*(\mathcal{T})$ with a lattice in $\mathfrak{g}$. The duality pairing between $\mathfrak{g}$ and $\mathfrak{g}^*$ induces a pairing $\varkappa$ between $X(\mathcal{T})$ and $X_*(\mathcal{T})$ which can be described intrinsically as follows. For given $\chi \in X(\mathcal{T})$ and $\psi \in X_*(\mathcal{T})$ we obtain an element $\chi \circ \psi$ in End $\mathbb{G}_m$ and, since End $\mathbb{G}_m$ is canonically isomorphic to $\mathbb{Z}$, we can take $\varkappa(\chi, \psi)$ to be the image of $\chi \circ \psi$ in $\mathbb{Z}$; then one easily verifies that the function $\varkappa$ has the properties of a pairing. The argument applies analogously in the case $G = \mathbb{G}_a$ if we define $l: X' = \mathrm{Hom}(G, \mathbb{G}_a) \to \mathfrak{g}^*$ by $l_\chi = d\chi$ and $\partial: X'_* \to \mathfrak{g}$ by $\partial_\psi = (d\psi)(d/d\chi)$ with χ a fixed generator of X' when this is viewed as the additive character group.

4.6 Subgroup varieties

In this section we determine the subgroup varieties of the group variety

$$G = \mathbb{G}_a \times \mathbb{G}_m^n .$$

A good reference here is Borel's book [51] which includes in particular the relevant theory relating to tori.

Proposition 4.3 *If $H \subseteq G$ is an algebraic subgroup then $H = H_a \times H_m$ where $H_a \subseteq \mathbb{G}_a$ and $H_m \subseteq \mathcal{T} = \mathbb{G}_m^n$ are algebraic subgroups.*

Proof. Let p_a and p_m be the restrictions to H of the projections $G \to \mathbb{G}_a$ and $G \to \mathbb{G}_m^n$ respectively. The algebraic subgroups $H_a = p_a^{-1}(0)$ and $H_m = p_m^{-1}(0)$ intersect only in 0 and so H contains the product $H_a \times H_m$. The quotient group given by

$$(\mathbb{G}_a \times \mathbb{G}_m^n)/(H_a \times H_m) \cong (\mathbb{G}_a/H_a) \times (\mathbb{G}_m/H_m)$$

contains the image H' of H and the homomorphisms p_a and p_m induce isomorphisms p'_a and p'_m from H' onto their images. This means that H' is the graph of an isomorphism between some algebraic subgroup of $\mathbb{G}_a$ and another of $\mathbb{G}_m^n$. As we noted above, such a homomorphism must be trivial. Hence we have $H' = 0$ and thus $H = H_a \times H_m$. □

The proposition shows that we need to determine subtori $\mathcal{S} \subseteq \mathcal{T}$ where $\mathcal{T}$ is the n-dimensional torus $\mathbb{G}_m^n$. In Section 4.5 we associated with a torus $\mathcal{T}$ of dimension n a (multiplicative) lattice $X(\mathcal{T})$, that is a free abelian group of rank n. Conversely we can associate a torus with a (multiplicative) lattice U of finite rank n. In fact, the group algebra $\mathbb{K}[U]$ of U is by definition the vector space with the elements of U as free generators and the algebra product of two elements u, v in U is the group element uv in U. The product can be extended linearly to $\mathbb{K}[U]$ and it defines the algebra structure. Then $\mathbb{K}[U]$ is the affine algebra of some torus $\mathcal{T}$ of dimension n. This construction gives a bijective correspondence between the class of tori and the class of (multiplicative) lattices of finite rank. Moreover, our discussion shows that the affine coordinate ring $\mathbb{K}[\mathcal{T}]$ of a torus $\mathcal{T}$ with character group $X(\mathcal{T})$ can be written canonically as $\mathbb{K}[X(\mathcal{T})]$, the algebra generated by the character group.

In a similar way, a homomorphism $\alpha: \mathcal{T}' \to \mathcal{T}$ from one torus into another defines a homomorphism $\alpha^*: X(\mathcal{T}) \to X(\mathcal{T}')$ between the corresponding character groups; it maps a character in $X(\mathcal{T})$ to its restriction to $\mathcal{T}'$. In the other direction, a group homomorphism from a lattice of finite rank to another determines uniquely a homomorphism between the corresponding group algebras and gives a unique homomorphism between the associated tori. Under the bijective correspondence a connected subtorus $\mathcal{S}$ of a given torus $\mathcal{T}$ determines a surjective homomorphism from $X(\mathcal{T})$ to $X(\mathcal{S})$. Now it is an established fact from group theory that a subgroup U of $X(\mathcal{T})$ is a direct factor if and only if the quotient group $X(\mathcal{T})/U$ is torsion free. Thus, if U is the kernel of the homomorphism above then it follows that U is a direct factor of $X(\mathcal{T})$.

We are now in a position to prove a lemma which is fundamental for transcendence theory over group varieties.

Lemma 4.4 *For any connected subtorus $\mathcal{S}$ of $\mathcal{T} = \mathbb{G}_m^n$ with codimension s there exist linear forms $L_1, \ldots, L_s$ with integer coefficients such that the Lie algebra of $\mathcal{S}$ is given by $L_1 = \cdots = L_s = 0$.*

Proof. Let $\mathfrak{g}$ be the Lie algebra of $\mathcal{T}$ and let $\mathfrak{g}^*$ be the dual space of $\mathfrak{g}$. Further let Y be the subgroup of $X(\mathcal{T})$ consisting of elements χ such that

$$\chi = 1$$

on $\mathcal{S}$. This is a set of generators for the ideal of $\mathcal{S}$. The homomorphism l introduced in Section 4.5 gives an embedding of Y in $\mathfrak{g}^*$ and the linear forms l_χ on $\mathfrak{g}$ for $\chi \in Y$ give equations $l_\chi = 0$ for the Lie algebra $\mathfrak{g}(\mathcal{S})$ of $\mathcal{S}$ in $\mathfrak{g}$. If $\chi_1, \ldots, \chi_n$ is a basis for $X(\mathcal{T})$ then the equations for $\mathcal{S}$ become

$$\chi^l = 1$$

with $\chi^l = \chi_1^{l(1)} \cdots \chi_n^{l(n)}$ in Y where l runs through a direct factor U of $\mathbb{Z}^n$ with rank s. We have

$$l_\chi = l(1)\, l_{\chi_1} + \cdots + l(n)\, l_{\chi_n}$$

and $\Delta\chi = l_\chi(\Delta)\chi$ for $\Delta \in \mathfrak{g}$. Thus the equations for the Lie algebra $\mathfrak{g}(\mathcal{S})$ can be written as $L = 0$ where

$$L = l(1)\, z_1 + \cdots + l(n)\, z_n\,;$$

here $z_1, \ldots, z_n$ are the coordinate functions of $\mathfrak{g}$ with respect to the basis $l_{\chi_1}, \ldots, l_{\chi_n}$ of $\mathfrak{g}^*$ and $l(j) \in \mathbb{Z}$. The lemma follows on taking a basis $\chi^{l_1}, \ldots, \chi^{l_s}$ for U and $L_1, \ldots, L_s$ as the corresponding linear form L. □

4.7 Geometry of Numbers

In this section we shall give some variations on the theme of Minkowski's theorem on successive minima in the Geometry of Numbers. We refer to Cassels' book [64] for basic definitions.

By a lattice $\Lambda \subset \mathbb{R}^n$ we mean a discrete subgroup of $\mathbb{R}^n$ which generates $\mathbb{R}^n$ over $\mathbb{R}$. Any lattice Λ can be written as

$$\Lambda = \mathbb{Z}\mathbf{a}_1 + \cdots + \mathbb{Z}\mathbf{a}_n$$

and the vectors $\mathbf{a}_1, \ldots, \mathbf{a}_n$ form a basis of $\mathbb{R}^n$. We associate with any such lattice Λ a real number $d(\Lambda)$, the determinant of Λ, defined as the absolute value of the determinant of the $n \times n$ matrix whose rows are the vectors $\mathbf{a}_1, \ldots, \mathbf{a}_n$. Clearly $d(\Lambda)$ depends only on Λ and not on the choice of basis. If one is given a set of lattice elements one can ask the question whether it is possible to extend the set to a basis for the lattice. The following lemma gives a necessary and sufficient condition under which this is possible.

Lemma 4.5 *Let $\mathbf{b}_1, \ldots, \mathbf{b}_m$ be elements of a lattice Λ, with $1 \leq m \leq n$, and let $\mathbf{c}_1, \ldots, \mathbf{c}_m$ be their coordinate vectors with respect to a basis $\mathbf{a}_1, \ldots, \mathbf{a}_n$ of Λ. Then $\mathbf{b}_1, \ldots, \mathbf{b}_m$ can be extended to a basis of Λ if and only if the determinant of the $m \times m$ minors of the $m \times n$ matrix with row vectors $\mathbf{c}_1, \ldots, \mathbf{c}_m$ have no common factor.*

Proof. See [64, Ch. I, §2, Lemma 2]. □

We recall that a set of vectors $\mathbf{b}_1, \ldots, \mathbf{b}_m$ as above is said to be primitive if the condition on the minors of the matrix attaching to $\mathbf{c}_1, \ldots, \mathbf{c}_m$ is satisfied. In view of Lemma 4.5, a primitive set of vectors $\mathbf{b}_1, \ldots, \mathbf{b}_m$ can be extended to a basis of Λ.

By a convex body B we mean a subset $B \subseteq \mathbb{R}^n$ such that for any $\mathbf{x}$, $\mathbf{y}$ in B and all real numbers t with $0 \leq t \leq 1$ the element $t\mathbf{x} + (1 - t)\mathbf{y}$ is in B. The convex body B is called symmetric if $-\mathbf{x}$ is in B whenever the vector $\mathbf{x}$ is in B. Let now B be a bounded convex symmetric body and let Λ be a lattice in $\mathbb{R}^n$. We define the kth successive minimum $\lambda_k = \lambda_k(B, \Lambda)$ to be the infimum of all real numbers $\lambda > 0$ such that λB contains k linearly independent lattice points. Then trivially we have

$$\lambda_1 \leq \lambda_2 \leq \cdots \leq \lambda_n .$$

The space $\mathbb{R}^n$ is equipped with the standard Lebesgue measure and one sees that every bounded convex body is Lebesgue measurable and possesses a volume $\operatorname{vol} B$. It is well known [64, Ch. VIII] that the successive minima $\lambda_1, \ldots, \lambda_n$ of a bounded symmetric convex body B with respect to a lattice Λ in $\mathbb{R}^n$ satisfy

$$(2^n/n!)d(\Lambda) \leq \lambda_1 \cdots \lambda_n \operatorname{vol} B \leq 2^n d(\Lambda) .$$

We next introduce a weighted volume for a submodule $\mathcal{M}$ over $\mathbb{Z}$ of the particular lattice $\Lambda = \mathbb{Z}^n$. Since Λ is a free $\mathbb{Z}$-module the same holds for $\mathcal{M}$ and we can choose a basis $\mathbf{m}_1, \ldots, \mathbf{m}_r$ for $\mathcal{M}$; here r is called the rank of $\mathcal{M}$. We denote by M the $r \times n$ matrix with row vectors $\mathbf{m}_1, \ldots, \mathbf{m}_r$ and we write $M_{j_1,\ldots,j_r}$ for the $r \times r$ minor corresponding to a set of indices $j_1, \ldots, j_r$ with $1 \le j_1 < \cdots < j_r \le n$. Then for any real vector $\mathbf{a} = (a_1, \ldots, a_n)$ with positive coordinates we define the weighted volume of $\mathcal{M}$ with respect to $\mathbf{a}$ by

$$\mathrm{vol}_{\mathbf{a}}(\mathcal{M}) = \max\left(a_{j_1} \cdots a_{j_r} \left|\det(M_{j_1,\ldots,j_r})\right|\right),$$

where the maximum is taken over all possible choices of $j_1, \ldots, j_r$. The definition does not depend on the choice of basis; in fact if the matrices N and $N_{j_1,\ldots,j_r}$ are obtained in the same way as M and $M_{j_1,\ldots,j_r}$ on starting with a different basis then we obtain $N = UM$ for some $r \times r$ matrix U with integral coefficients and determinant ± 1. Further the volume of $\mathcal{M}$ is independent of the choice of basis for we have $N_{j_1,\ldots,j_r} = UM_{j_1,\ldots,j_r}$ and therefore $\left|\det\left(N_{j_1,\ldots,j_r}\right)\right| = \left|\det(M_{j_1,\ldots,j_r})\right|$.

We now show that there exists a sublattice of $\mathcal{M}$ that is generated by elements possessing a property of Minkowski type in terms of the distance function

$$G_{\mathbf{a}}(\mathbf{x}) = a_1|x_1| + \cdots + a_n|x_n|$$

which is defined with respect to a vector $\mathbf{a} = (a_1, \ldots, a_n)$ with positive real coordinates.

Lemma 4.6 *For any submodule $\mathcal{M}$ of $\mathbb{Z}^n$ of rank r there exist linearly independent elements $\mathbf{y}_1, \ldots, \mathbf{y}_r$ of $\mathcal{M}$ such that*

$$G_{\mathbf{a}}(\mathbf{y}_1) \cdots G_{\mathbf{a}}(\mathbf{y}_r) \le (n+1-r)^r r! \,\mathrm{vol}_{\mathbf{a}}(\mathcal{M}) .$$

Proof. Let $\mathbf{m}_1, \ldots, \mathbf{m}_r$ be a basis for $\mathcal{M}$ and let M be the matrix with rows $\mathbf{m}_1, \ldots, \mathbf{m}_r$. We fix integers $j_1, \ldots, j_r$ with

$$\mathrm{vol}_{\mathbf{a}}(\mathcal{M}) = a_{j_1} \cdots a_{j_r} \left|\det(M_{j_1,\ldots,j_r})\right| ,$$

and define the projection $\pi : \mathbb{R}^n \to \mathbb{R}^r$ by sending $\mathbf{x} = (x_1, \ldots, x_n)$ to $\mathbf{x}' = \pi(\mathbf{x}) = (x_{j_1}, \ldots, x_{j_r})$. The distance functions in $\mathbb{R}^r$ are then given by

$$G'_{\mathbf{a}}(\mathbf{x}') = a_{j_1}\left|x_{j_1}\right| + \cdots + a_{j_r}\left|x_{j_r}\right| .$$

Since the elements $\mathbf{m}_1, \ldots, \mathbf{m}_r$ form a basis of $\mathcal{M}$ so that $\det(M_{j_1,\ldots,j_r}) \neq 0$, we can express every vector $\mathbf{x}' \in \mathbb{R}^r$ uniquely as

$$\mathbf{x}' = \xi_1 \pi(\mathbf{m}_1) + \cdots + \xi_r \pi(\mathbf{m}_r)$$

with real coefficients $\xi_1, \ldots, \xi_r$. Hence we can define a mapping $\sigma : \mathbb{R}^r \to \mathbb{R}^n$ by

$$\mathbf{x} = \sigma(\mathbf{x}') = \xi_1 \mathbf{m}_1 + \cdots + \xi_r \mathbf{m}_r .$$

This is a homomorphism with the property that $\pi \circ \sigma = \mathrm{id}$ and for $\mathbf{x}' \in \pi(\mathcal{M})$ we have $\sigma(\mathbf{x}') \in \mathcal{M}$.

It will suffice now to show that

$$G_{\mathbf{a}}(\sigma(\mathbf{x}')) \leq (n+1-r) G'_{\mathbf{a}}(\mathbf{x}') .$$

For let $\lambda_1, \ldots, \lambda_r$ be the successive minima with respect to the lattice $\pi(M)$ of the convex body B' given by $G'_{\mathbf{a}}(\mathbf{x}') < 1$. The volume $\operatorname{vol} B'$ of B' is calculated as $(2^r/r!)\left(a_{j_1} \cdots a_{j_r}\right)^{-1}$ and the determinant $d(\pi(\mathcal{M}))$ of $\pi(\mathcal{M})$ as $\left|\det\left(M_{j_1,\ldots,j_r}\right)\right|$. Thus by Minkowski's theorem we have

$$\lambda_1 \cdots \lambda_r \leq r! \operatorname{vol}_{\mathbf{a}}(\mathcal{M}) .$$

By the definition of the successive minima there exist linearly independent elements $\mathbf{y}'_1, \ldots, \mathbf{y}'_r$ in $\pi(\mathcal{M})$ with $G'_{\mathbf{a}}(\mathbf{y}'_j) = \lambda_j$. We put $\mathbf{y}_j = \sigma(\mathbf{y}'_j)$ for $1 \leq j \leq r$. Then

$$G_{\mathbf{a}}(\mathbf{y}_j) \leq (n+1-r) G'_{\mathbf{a}}\left(\mathbf{y}'_j\right) .$$

Hence

$$G_{\mathbf{a}}(\mathbf{y}_1) \cdots G_{\mathbf{a}}(\mathbf{y}_r) \leq (n+1-r)^r \lambda_1 \cdots \lambda_r$$

and the inequality for $\lambda_1 \cdots \lambda_r$ yields the stated result.

It remains to prove the inequality relating $G_{\mathbf{a}}$ and $G'_{\mathbf{a}}$. We denote the columns of the matrix M by $\mu_1, \ldots, \mu_n$. They generate a vector space over the rationals of dimension r and the column vectors $\mu_{j_1}, \ldots, \mu_{j_r}$ are linearly independent. Hence we can write μ_ϱ for $1 \leq \varrho \leq n$ as a linear combination

$$\mu_\varrho = \gamma_{1\varrho} \mu_{j_1} + \cdots + \gamma_{r\varrho} \mu_{j_r}$$

with rational coefficients γ_{jk}. These can be expressed as quotients of determinants of minors of the matrix M. In fact we have

$$\gamma_{\sigma\varrho} = \pm \det\left(M_{j_\sigma}\right) / \det\left(M_{j_1,\ldots,j_r}\right)$$

where M_{j_σ} denotes the $r \times r$ matrix $M_{j_1,\ldots,j_r}$ with the column μ_{j_σ} replaced by μ_ϱ. By the definition of $j_1, \ldots, j_r$ we see that the absolute value of this quotient is at most equal to a_{j_σ}/a_ϱ; hence we obtain

$$a_\varrho |\gamma_{\sigma\varrho}| \leq a_{j_\sigma} .$$

If we write $\sigma(\mathbf{x}') = \mathbf{x} = (x_1, \ldots, x_n)$ and $\mathbf{m}_j = \left(m_{j_1}, \ldots, m_{j_n}\right)$ then

$$\begin{aligned} x_k &= \sum_{\sigma=1}^{r} \xi_\sigma m_{\sigma k} = \sum_{\sigma=1}^{r} \xi_\sigma \sum_{\varrho=1}^{r} \gamma_{\varrho k} m_{\sigma j_\varrho} \\ &= \sum_{\varrho=1}^{r} \gamma_{\varrho k} \sum_{\sigma=1}^{r} \xi_\sigma m_{\sigma j_\varrho} = \sum_{\varrho=1}^{r} \gamma_{\varrho k} x_{j_\varrho}. \end{aligned}$$

On multiplying by a_k and using the above inequality we get

$$a_k |x_k| \leq \sum_{\varrho=1}^{r} a_{j_\varrho} |x_{j_\varrho}| .$$

The right-hand side is $G'_{\mathbf{a}}(\mathbf{x}')$. Clearly by the definition of $G_{\mathbf{a}}(\mathbf{x})$ we have

$$G_{\mathbf{a}}(\mathbf{x}) = G'_{\mathbf{a}}(\mathbf{x}') + \sum_k a_k |x_k|$$

where the sum is taken over all $k \neq j_1, \ldots, j_r$, and there are exactly $n - r$ terms in the latter sum each of which is bounded by $G'_{\mathbf{a}}(\mathbf{x}')$. Hence finally

$$G_{\mathbf{a}}(\mathbf{x}) \leq (n + 1 - r) G'_{\mathbf{a}}(\mathbf{x}')$$

as stated. □

Lemma 4.6 shows that there exists a sublattice for $\mathcal{M}$. Incidentally, Kunrui Yu has pointed out to us that, by [176], the constant $(n + 1 - r)^r r!$ here can be improved to $n(n-1)\cdots(n-r+1)$ and Masser has remarked that another approach, using the Cauchy–Binet identity instead of projections, gives the strange coefficient

$$\left(\frac{4n}{\pi}\right)^{\frac{1}{2}r}\sqrt{\binom{n}{r}}\,\Gamma\left(\frac{r}{2}+1\right)$$

which seems numerically better if $r < n$. However, it will be essential in the sequel to exhibit a basis. This can be done although at some cost (cf. [64, Ch. V, Lemma 8]). Namely, with the same hypothesis as Lemma 4.6, we prove the following.

Lemma 4.7 *There exists a basis* $\mathbf{x}_1, \dots, \mathbf{x}_r$ *of* $\mathcal{M}$ *such that*

$$G_{\mathbf{a}}(\mathbf{x}_1)\cdots G_{\mathbf{a}}(\mathbf{x}_r) \le (n+1-r)^r (r!)^2 \left(\tfrac{1}{2}\right)^{r-1} \mathrm{vol}_{\mathbf{a}}(\mathcal{M})\,.$$

Proof. Let $\mathbf{y}_1, \dots, \mathbf{y}_r$ be the vectors given by Lemma 4.6 such that $G_{\mathbf{a}}(\mathbf{y}_j) \le (n+1-r)\lambda_j$. We denote by $\mathcal{N}_j$ the vector space generated over the rationals by $\mathbf{y}_1, \dots, \mathbf{y}_j$. On intersecting $\mathcal{N}_j$ with $\mathcal{M}$ we obtain a submodule $\mathcal{L}_j$ of the lattice $\mathcal{M}$ of rank j. We proceed to construct inductively vectors $\mathbf{x}'_1, \dots, \mathbf{x}'_r$ such that $\mathbf{x}'_1, \dots, \mathbf{x}'_j$ is a basis for $\mathcal{L}_j$.

By definition we have $\mathcal{L}_1 = \mathbb{Z}\beta_1\mathbf{y}_1$ for some rational β_1 and we put $\mathbf{x}'_1 = \beta_1\mathbf{y}_1$. Now we assume that $\mathbf{x}'_1, \dots, \mathbf{x}'_{j-1}$ form a basis for $\mathcal{L}_{j-1}$ for each j with $1 < j \le r$. We choose for $\mathbf{x}'_j$ a vector in $\mathcal{L}_j$ which has minimal positive distance to $\mathcal{N}_{j-1}$ with respect to $G_{\mathbf{a}}$. Then $\mathbf{x}'_1, \dots, \mathbf{x}'_{j-1}, \mathbf{x}'_j$ constitute a basis for $\mathcal{L}_j$. In fact let $\mathbf{x} = \delta_1\mathbf{x}'_1 + \cdots + \delta_j\mathbf{x}'_j$ be in $\mathcal{L}_j$ where the δ_i are rational. Then we define

$$\mathbf{x}' = \{\delta_1\}\mathbf{x}'_1 + \cdots + \{\delta_j\}\mathbf{x}'_j \in \mathcal{L}_j\,,$$

where $\{\delta\}$ denotes the fractional part of δ so that $0 \le \{\delta\} < 1$. The distance of $\mathbf{x}'$ to $\mathcal{N}_{j-1}$ is $\{\delta_j\}$ times the distance of $\mathbf{x}'_j$ to $\mathcal{L}_{j-1}$ and so, by the minimality of the latter, we conclude that $\{\delta_j\}$ has to be zero, that is $\delta_j \in \mathbb{Z}$. By induction we obtain $\delta_i \in \mathbb{Z}$ for $i < j$, whence $\mathbf{x}'_1, \dots, \mathbf{x}'_j$ is a basis for the $\mathbb{Z}$-module $\mathcal{L}_j$.

Let $\mathcal{M}_j$ be the lattice generated by $\mathbf{x}'_1, \dots, \mathbf{x}'_j, \mathbf{y}_{j+1}, \dots, \mathbf{y}_r$. Then by the construction of $\mathbf{x}'_1, \dots, \mathbf{x}'_r$ we have $\mathcal{M}_{j-1} \subseteq \mathcal{M}_j$ and this gives a filtration

$$\mathcal{M} = \mathcal{M}_r \supseteq \mathcal{M}_{r-1} \supseteq \cdots \supseteq \mathcal{M}_1 \supseteq \mathcal{M}_0\,.$$

We have

$$\mathbf{x}'_j = \beta_1\mathbf{y}_1 + \cdots + \beta_{j-1}\mathbf{y}_{j-1} + \beta_j\mathbf{y}_j$$

with rationals $\beta_1, \ldots, \beta_j$, and the distance of $\mathbf{x}'_j$ to $\mathcal{N}_{j-1}$ is equal to $|\beta_j|$ times the distance of $\mathbf{y}_j$ to $\mathcal{N}_{j-1}$. It follows by the minimal assumption on $\mathbf{x}'_j$ that $|\beta_j| \leq 1$. In fact it is easily seen that $\left|\beta_j\right|$ has to be $1/e_j$ for any j where $e_j \in \mathbb{N}$ is the index of $\mathcal{M}_{j-1}$ in $\mathcal{M}_j$ but we do not need to use the observation here.

We now set $\mathbf{x}_1 = \mathbf{x}'_1$ and for $1 < j \leq r$ we put $\mathbf{x}_j = \mathbf{y}_j$ if $|\beta_j| = 1$ and

$$\mathbf{x}_j = (\beta_1 - b_1)\mathbf{y}_1 + \cdots + \left(\beta_{j-1} - b_{j-1}\right)\mathbf{y}_{j-1} + \beta_j\mathbf{y}_j$$

if $|\beta_j| < 1$, where the b_i are integers such that $|\beta_i - b_i| \leq \frac{1}{2}$. Clearly $\mathbf{x}_1, \ldots, \mathbf{x}_r$ is again a basis of $\mathcal{M}$. Since $\lambda_1 \leq \lambda_2 \leq \cdots \leq \lambda_r$, we have the estimates

$$G_{\mathbf{a}}(\mathbf{x}_1) \leq G_{\mathbf{a}}(\mathbf{y}_1) \leq (n+1-r)\lambda_1$$

and

$$G_{\mathbf{a}}\left(\mathbf{x}_j\right) \leq \tfrac{1}{2}\sum_{v=1}^{j} G_{\mathbf{a}}(\mathbf{y}_v) \leq \tfrac{1}{2}(n+1-r)j\lambda_j \quad (1 < j \leq r).$$

These give

$$G_{\mathbf{a}}(\mathbf{x}_1)\cdots G_{\mathbf{a}}(\mathbf{x}_r) \leq (n+1-r)^r r!\left(\tfrac{1}{2}\right)^{r-1}\lambda_1\lambda_2\cdots\lambda_r$$

and Lemma 4.7 follows from our earlier upper bound for $\lambda_1 \cdots \lambda_r$. □

5
Multiplicity estimates

5.1 Hilbert functions in degree theory

We now come to the preliminary results on degree theory needed for the proof of the fundamental multiplicity estimates in Section 5.5 and we begin, in this section, with an account of the theory of Hilbert functions. We shall follow as much as possible the exposition given in sections 10 and 11 of the book of Atiyah and Macdonald [9].

Let I be the set of integers j with $0 \leq j \leq n$ and let $\Gamma = \mathbb{N}^I$ be the set of functions on I with values in $\mathbb{N}$ which is a free monoid with $n+1$ generators. An I-graded ring is a ring A together with a family of subgroups $(A_\gamma)_{\gamma \in \Gamma}$ of the additive group of A such that $A = \bigoplus_{\gamma \in \Gamma} A_\gamma$ and $A_{\gamma_1} A_{\gamma_2} \subseteq A_{\gamma_1+\gamma_2}$ for all $\gamma_1, \gamma_2 \in \Gamma$. In particular A_0 is a subring of A and each A_γ is a module over A_0. In our subsequent application of the theory we shall take for A the polynomial ring $R = \mathbb{K}[X_0, Y_0; \ldots; X_n, Y_n]$ in $2(n+1)$ variables over a field $\mathbb{K}$ and for A_γ the subgroup R_γ of R generated by all monomials $M(\gamma_1, \gamma_2) = X_0^{\gamma_1(0)} Y_0^{\gamma_2(0)} \cdots X_n^{\gamma_1(n)} Y_n^{\gamma_2(n)}$ with $\gamma_1 + \gamma_2 = \gamma$. Hence R_γ will then consist of all polynomials which are homogeneous in the pairs of variables X_i, Y_i of degree $\gamma(i)$ for $i = 0, 1, \ldots, n$.

A graded A-module, where A is a graded ring, is defined as a module M together with a family of subgroups $(M_\gamma)_{\gamma \in \Gamma}$ of the additive group of M such that $M = \bigoplus_{\gamma \in \Gamma} M_\gamma$ and $A_{\gamma_1} M_{\gamma_2} \subseteq M_{\gamma_1+\gamma_2}$ for all $\gamma_1, \gamma_2 \in \Gamma$. Again this implies that each M_γ is an A_0-module. Elements of M_γ are called homogeneous of degree γ. Further, each $m \in M$ can be expressed uniquely as a finite sum $\sum m_\gamma$ for $m_\gamma \in M_\gamma$ and m_γ is called the homogeneous component of m.

Let now A be a Noetherian graded ring. Then also A_0 is Noetherian and A is generated over A_0 by homogeneous elements $a_1, \ldots, a_s$ of non-zero degrees $d_1, \ldots, d_s \in \Gamma$. We take M to be a finitely generated graded A-module whence each M_γ is a finitely generated A_0-module. We fix some additive function λ on the class of finitely generated A_0-modules, that is an integer-valued function λ which satisfies $\lambda(M) = \lambda(M') + \lambda(M'')$ if $0 \to M \to M' \to M'' \to 0$ is an exact sequence. The Poincaré series of M (with respect to λ) is the generating function

$$P(M,t) = \sum_{\gamma \in \Gamma} \lambda(M_\gamma)\, t^\gamma .$$

Here t denotes a set of variables $t_0, \ldots, t_n$ and $t^\gamma = t_0^{\gamma(0)} \cdots t_n^{\gamma(n)}$. Then

$$P(M,t) = f(t) \,/\, \prod_{i=1}^{s} (1 - t^{d_i}),$$

where $f(t)$ is a polynomial with integer coefficients, and so $P(M,t)$ is a rational function of t. This is a generalisation of a famous theorem of Hilbert. The proof of the result is formally the same as the proof given in [9] in the case when I consists only of one element and we shall not repeat the arguments here. The generating function $P(M,t)$ encodes all the relevant information about the module M.

We consider now the case when the functions d_i, $1 \le i \le s$, are taken from the set of functions $\varepsilon_i \in \Gamma$, $0 \le i \le n$, which are defined by $\varepsilon_i(j) = \delta_{i,j}$ where $\delta_{i,j}$ denotes the usual Kronecker delta function. Then the denominator of the Poincaré series takes the form $\prod_{i=0}^{n} (1 - t^{\varepsilon_i})^{m(i)}$ with $m \in \Gamma$ and with $n \ge 0$. The latter product can be developed into a series of the shape $\sum_{\varkappa \in \Gamma} b_\varkappa t^\varkappa$ where $b_\varkappa$ is given by

$$\binom{\varkappa(0) + m(0) - 1}{m(0) - 1} \cdots \binom{\varkappa(n) + m(n) - 1}{m(n) - 1}.$$

Hence, if $f(t) = \sum_{\mu \in \Gamma} f_\mu t^\mu$ is a polynomial then

$$\lambda(M_\gamma) = \sum_{0 \le \mu \le \gamma} b_{\gamma - \mu} f_\mu .$$

Now, the function $\lambda(M_\gamma)$ is a polynomial in $\gamma(0), \ldots, \gamma(n)$, usually called the Hilbert function (or polynomial) $\chi_M(\gamma)$ of M with respect to λ. The degree of $\chi_M(\gamma)$ in the variable $\gamma(i)$ is at most $m(i) - 1$, by construction, and it can be written as a linear combination

$$\chi_M(\gamma) = \sum_{k \geq 0} \chi_{M,k}(\gamma)$$

of homogeneous polynomials $\chi_{M,k}(\gamma)$ of degree k (in $\mathbb{N}$). We denote by $d(M)$ the smallest integer l such that $\chi_{M,k}(\gamma) = 0$ for all $k > l$. The term $\chi_{M,d(M)}(\gamma)$ of highest degree $d(M)$ contains all the information relevant to us; if the function λ is non-negative then this term is non-negative and can be written as

$$\sum_{\substack{j(0)+\cdots+j(n)=d(M) \\ j \leq m-1}} \lambda_j(M)\, \gamma^j / j!,$$

where $j! = j(0)! \cdots j(n)!$ and $\gamma^j = \gamma(0)^{j(0)} \cdots \gamma(n)^{j(n)}$. We obtain for $j \in \Gamma$ additive functions $\lambda_j \geq 0$ on the class of finitely generated modules over A_0. In summary, we associate with an additive function λ on the class of finitely generated A_0-modules a function $d \neq 0$ with values in $\mathbb{N}$ and a family of additive functions $\lambda_j \colon M \mapsto \lambda_j(M)$; the values of the latter are not all zero if $j(0) + \cdots + j(n) = d(M), j \leq m - 1$, and are zero otherwise. If λ takes integer values then so do all the λ_j and these are not all zero if and only if λ is non-zero. The function d is equal to $-\infty$ if $M = 0$.

We now return to the example introduced above and we take here as generators $a_1, \ldots, a_s$ the variables X_i and Y_i for $0 \leq i \leq n$ with degree ε_i. Then the family of d_i, $1 \leq i \leq s$, contains exactly two of the functions ε_i with $0 \leq i \leq n$. Thus we have $m(i) \leq 2$ for each $i = 0, 1, \ldots, n$ and furthermore we have $\lambda_j = 0$ when j does not satisfy $j \leq 1$. In other words the support of the function $\lambda \colon j \mapsto \lambda_j$ is contained in the subset of Γ which consists of the functions j which have values in the set $\{0, 1\}$. We apply our theory to quotients R/I of R by some homogeneous ideal I, that is an ideal of R given by the direct sum $\bigoplus_\omega I_\omega$ of $I_\omega = I \cap R_\omega$. It is now easy to verify the following.

Proposition 5.1 *Let $I, J \subseteq R$ be homogeneous ideals and let $\Phi \in R$ be a homogeneous element. Then we have*

$$P(R/(I \cap J), t) = P(R/I, t) + P(R/J, t) - P(R/(I + J), t) \quad (5.1)$$

and

$$P(R/\Phi I, t) = t^{\deg \Phi} P(R/I, t) + P(R/(\Phi), t). \quad (5.2)$$

Proof. We use the fact that the Poincaré series is additive with respect to the module M and that the sequences

$$0 \longrightarrow (I + J)/I \longrightarrow R/I \longrightarrow R/(I + J) \longrightarrow 0$$

and

$$0 \longrightarrow J/(I \cap J) \longrightarrow R/(I \cap J) \longrightarrow R/J \longrightarrow 0$$

are both exact. Since $(I+J)/I \cong J/(I \cap J)$ the first part of the proposition follows. For the second part we observe that multiplication by Φ gives an endomorphism of the module R of degree $\deg \Phi$ and induces an exact sequence

$$0 \longrightarrow R/I \stackrel{\Phi}{\longrightarrow} R/\Phi I \longrightarrow R/(\Phi) \longrightarrow 0.$$

The result follows on using again the additivity of the Poincaré series with respect to M and incorporating a shift in the degree. □

We now need a definition. If $I \subseteq R$ is an ideal and $\Phi \in R$ we have $I \cap (\Phi) = \Phi(I : \Phi)$; if now $I : \Phi = I$ then we say that I and Φ are coprime.

Corollary 5.2 *Suppose that I and Φ are coprime. Then*

$$P(R/(\Phi, I), t) = P(R/I, t) - t^{\deg \Phi} P(R/I, t).$$

Proof. We combine (5.1) with (5.2) and use coprimality. □

Let $\mathfrak{Q} \subseteq R$ be a primary ideal and let $\mathfrak{P}$ be its radical. Then there exists a composition series $\mathfrak{P} = \mathfrak{Q}_0 \supset \mathfrak{Q}_1 \supset \cdots \supset \mathfrak{Q}_{l-1} \supset \mathfrak{Q}$ of length l with $\mathfrak{Q}_i/\mathfrak{Q}_{i+1}$ simple for $0 \le i \le l - 1$. Since the sequence

$$0 \longrightarrow \mathfrak{Q}_i/\mathfrak{Q}_{i+1} \longrightarrow R/\mathfrak{Q}_{i+1} \longrightarrow R/\mathfrak{Q}_i \longrightarrow 0$$

is exact and since the Poincaré series is additive we get

$$P(R/\mathfrak{Q}_i, t) = P(R/\mathfrak{Q}_{i+1}, t) - P(\mathfrak{Q}_i/\mathfrak{Q}_{i+1}, t)$$

and, on taking the sum over all i, this gives

$$P(\mathfrak{P}/\mathfrak{Q}, t) = \sum_{i=0}^{l-1} P(\mathfrak{Q}_i/\mathfrak{Q}_{i+1}, t).$$

Now each of the modules $\mathfrak{Q}_i/\mathfrak{Q}_{i+1}$ is a simple $R/\mathfrak{P}$-module and therefore isomorphic to $R/\mathfrak{P}$. It follows that

$$P(R/\mathfrak{Q}, t) = l(\mathfrak{Q})\, P(R/\mathfrak{P}, t),$$

where $l(\mathfrak{Q})$ denotes the length of the primary ideal $\mathfrak{Q}$.

5.2 Differential length

The concept of differential length is an essential ingredient of the degree theory which we shall utilise in the next section and which leads to the basic result on multiplicity estimates on group varieties. It occurred first in published form in Wüstholz [263]. The main result (5.4) that we establish here furnishes a lower bound for the length of a primary component of a particular polynomial ideal.

We consider the affine algebra $\mathbb{K}[G]$ of the algebraic group G introduced at the end of Section 4.3 and we fix a subspace $\mathfrak{h}$ of the Lie algebra of G. Let $\Delta_1, \ldots, \Delta_h$ be any basis for $\mathfrak{h}$ and let $\mathfrak{P} = (P_1, \ldots, P_k)$ be a prime ideal of $\mathbb{K}[G]$. We denote by $J(\mathfrak{P}, \mathfrak{h})$ the Jacobian matrix

$$J(\mathfrak{P}, \mathfrak{h}) = \begin{pmatrix} \Delta_1 P_1 & \cdots & \Delta_1 P_k \\ \vdots & & \vdots \\ \Delta_h P_1 & \cdots & \Delta_h P_k \end{pmatrix}$$

and by $\varrho(\mathfrak{P}, \mathfrak{h})$, briefly $\varrho(\mathfrak{P})$, its rank modulo $\mathfrak{P}$; we shall call this the Jacobi rank. The latter is obviously independent of the choice of basis of $\mathfrak{h}$ and of the generators P_i of $\mathfrak{P}$. For integers $t \geq 0$ we further denote by

$\Omega(t)$ the set of functions $\tau\colon \{1,\ldots,h\} \to \mathbb{N}$ such that $\tau(1)+\cdots+\tau(h) \leq t$ and we put

$$\Delta(\tau) = \Delta_1^{\tau(1)} \cdots \Delta_h^{\tau(h)}.$$

Let I be an ideal of R and let $I^{(t)}$ be the ideal

$$I^{(t)} = \{Q \in R;\ \forall \tau \in \Omega(t)\colon \Delta(\tau)\, Q \in \sqrt{I}\}, \tag{5.3}$$

where $\sqrt{I}$ denotes the radical of I. Clearly $(I \cap J)^{(t)} = I^{(t)} \cap J^{(t)}$ for all ideals I and J in R. Further, if I is a prime ideal $\mathfrak{P}$ then $\mathfrak{P}^{(t)}$ is primary with $\mathfrak{P}$ as associated prime. Furthermore, if $\mathfrak{P}$ is a prime ideal of R, we have, for integers $t \geq 0$,

$$l(\mathfrak{P}^{(t)}) \geq \binom{t + \varrho(\mathfrak{P})}{\varrho(\mathfrak{P})}. \tag{5.4}$$

In fact let $x_1, \ldots, x_\varrho$, where $\varrho = \varrho(\mathfrak{P})$, be any subset of a set of generators of $\mathfrak{P}$ such that the corresponding minor of the Jacobian has determinant not in $\mathfrak{P}$. Then the ideal generated by $\mathfrak{P}^{(t)}$ together with the elements $x_1^{\tau(1)} \cdots x_\varrho^{\tau(\varrho)}$ with $1 \leq \tau(1) + \cdots + \tau(\varrho) \leq t$ has a uniquely determined isolated primary component $\mathfrak{Q}$ with associated prime $\mathfrak{P}$ which satisfies

$$\mathfrak{P} \supseteq \mathfrak{Q} \supseteq \mathfrak{P}^{(t)}.$$

By a construction based on classical commutative algebra involving, in particular, systems of parameters, there exists a composition series $\mathfrak{Q} = \mathfrak{Q}_0 \supset \mathfrak{Q}_1 \supset \cdots \supset \mathfrak{Q}_{l-1} \supset \mathfrak{P}^{(t)}$ of length l, where l is at least equal to the binomial on the right of (5.4); this result forms a major constituent of Wüstholz [263] (see §3 therein). The desired inequality now follows on noting that the composition series can be extended to give one which begins with $\mathfrak{P}$ and therefore has length at least l.

In the main application to logarithmic forms one has $G = \mathbb{G}_a \times \mathbb{G}_m^n$ and $h = n$. We have given a wider discussion, however, since we have in mind applications to more general groups with different h. In particular, this applies to studies on Lindemann's theorem for abelian varieties as we shall discuss in Section 6.5.

5.3 Algebraic degree theory

In this section we shall develop a degree theory on the affine algebra of the algebraic group G of Section 5.2. Degree theory first entered into transcendence in the context of algebraic groups through the work of Masser and Wüstholz [164, 165]. Here the degrees of the relevant polynomials are the same but Masser and Wüstholz [166] needed results with polynomials of different degrees and they were thus led to multidegree theory. The basic Lemma 5.3, which we quote below, is that of Wüstholz [262] but we give here a new proof. It utilises Poincaré series rather than the method of calculating dimensions in the classical manner of Gröbner and van der Waerden.

We consider the ring of regular functions on G, that is the algebra $\mathbb{K}[G]$. For integers k with $0 \le k \le n+1$, we denote by $\Gamma(k) \subseteq \Gamma$ the set of functions $j\colon \{0, \ldots, n\} \to \{0, 1\}$ with function values $j(0), \ldots, j(n)$ satisfying $j(0) + \cdots + j(n) = k$. We now establish the following fundamental lemma.

Lemma 5.3 *For each k with $1 \le k \le n+1$ and for each $j \in \Gamma(k)$ there exists a non-vanishing function δ_j from the set of ideals of $\mathbb{K}[G]$ into the set of non-negative integers which has the following properties.*

(i) *Let I and J be ideals of the same rank. Then $I \subseteq J$ implies that $\delta_j(I) \ge \delta_j(J)$. Furthermore we have $\delta_j(I \cap J) \le \delta_j(I) + \delta_j(J)$ with equality if and only if the rank of $I + J$ is strictly greater than the rank of I and J.*

(ii) *We have, with the notation of Section 5.2,*

$$\delta_j(\mathfrak{P}^{(t)}) \ge \delta_j(\mathfrak{P}) \binom{t + \varrho(\mathfrak{P})}{\varrho(\mathfrak{P})}.$$

(iii) *Suppose that the ideal I has rank r and is generated by polynomials with degrees at most D_i in the variables X_i. Then, for $j \in \Gamma(r)$, we have*

$$\delta_j(I) \le r!\, D_0^{j(0)} \cdots D_n^{j(n)}.$$

(iv) *If $\mathfrak{M}$ is a maximal ideal in $\mathbb{K}[G]$ then $\delta_j(\mathfrak{M}) = 1$ if $j(0) + \cdots + j(n) = n+1$, and $\delta_j(\mathfrak{M}) = 0$ otherwise.*

We note that $\mathbb{K}[G]$ is the localisation of the ring of polynomials $\mathbb{K}[X_0, \ldots, X_n] \subseteq \mathbb{K}[G]$ with respect to the multiplicative monoid generated by the variables X_j with $1 \leq j \leq n$. The maps $I \mapsto I^e$ and $J \mapsto J^c$, where $I^e = I\,\mathbb{K}[G]$ and $J^c = J \cap \mathbb{K}[X_0, \ldots, X_n]$, called the extension and contraction mappings respectively, define a correspondence between the set of ideals in $\mathbb{K}[X_0, \ldots, X_n]$ and those in $\mathbb{K}[G]$. For the basic properties of the extension and contraction mappings adapted to our special situation we refer to Zariski and Samuel [270, I, Ch. IV, §10]. If R is the ring of polynomials introduced in Section 5.1 then $\mathbb{K}[X_0, \ldots, X_n]$ can be embedded into R by mapping a polynomial P of degree $d \in \Gamma$ to the homogeneous polynomial $P^h(X_0, Y_0; \ldots; X_n, Y_n)$ defined by $Y_0^{d(0)} \cdots Y_n^{d(n)} P(X_0/Y_0, \ldots, X_n/Y_n)$. We associate with an ideal $J \subseteq \mathbb{K}[X_0, \ldots, X_n]$ the homogeneous ideal $J^h \subseteq R$ defined as the ideal generated by all polynomials $Y_0^{m(0)} \cdots Y_n^{m(n)} P^h$ with $P \in J$ and $m \in \Gamma$. Conversely, if $I \subseteq R$ is a homogeneous ideal then we define I^a to be the ideal in R given by the set of polynomials P^a of the form $P(X_0, 1; \ldots; X_n, 1)$ with some homogeneous polynomial P. The correspondences $I \mapsto I^h$ and $J \mapsto J^a$ are almost inverse to each other; for details we refer to Zariski and Samuel [270, II, Ch. VII, §5].

Proof of Lemma 5.3. We take for ϱ the Jacobi rank $\varrho(\mathfrak{P})$ introduced in Section 5.2. The functions δ_j will be defined using the functions λ_j introduced in Section 5.1. For this we have to fix the additive function λ and here we take the function which associates to a vector space its dimension. Let $I \subseteq \mathbb{K}[G]$ be an ideal and let r denote the rank of I. In the case when $j(0) + \cdots + j(n) \neq r$ we take $\delta_j(I) = 0$. Otherwise we define $\delta_j(I)$ as $\lambda_{1-j}(R/(I^c)^h)$ as we may since $(I^c)^h$ is a homogeneous ideal in R. It remains to verify (i)–(iv).

Since the module $R/(J^c)^h$ is a quotient of the module $R/(I^c)^h$, the first statement in (i) follows from the additivity of the Poincaré series and the definition of δ_j; the second is a consequence of (5.1). For the proof of (ii) we apply (5.4) to get the result.

To establish (iii), let Σ be a set of generators of I and let $V \subseteq \mathbb{K}[X_0, \ldots, X_n]$ be the vector space generated by Σ. We construct, for $k = 1, \ldots, r$, a sequence of subspaces $W_k \subseteq V$ with dimension k such that, if $I_k \subseteq \mathbb{K}[G]$ is the ideal generated by W_k, then I_k has rank k. We have $I_{k+1} = (I_k, v)$ for some $v \neq 0$ in W_{k+1} coprime to I_k, whence

$$({I_{k+1}}^c)^h = ((I_k, v)^c)^h \supseteq (({I_k}^c) + (vR)^c)^h = ({I_k}^c)^h + ((vR)^c)^h.$$

We apply Corollary 5.2 to $({I_k}^c)^h$ and $(\Phi) = ((vR)^c)^h$, compare coefficients and deduce that for $j \in \Gamma(k+1)$ we have

$$\delta_j(I_{k+1}) \leq \sum \delta_{j-\varepsilon}(I_k)\delta_\varepsilon(v),$$

where the sum is over all $\varepsilon \in \Gamma(1)$ with $\varepsilon \leq j$. Taking $k = r$ we find recursively that, for $e_i \in \Gamma(1)$,

$$\delta_j(I_r) \leq \sum_{e_1+\cdots+e_r=j} \delta_{e_1}(v_1)\cdots\delta_{e_r}(v_r).$$

Now (iii) follows on observing that

$$\delta_{e_1}(v_1)\cdots\delta_{e_r}(v_r) \leq D_0^{j(0)}\cdots D_n^{j(n)}$$

and that the latter product with $e_1+\cdots+e_r = j$ is invariant under the action of the symmetric group in r elements.

It remains to construct inductively the sequence of ideals I_k. For $k = 1$ we take any non-zero $v_1 \in V$ and then the space $W_1 = \mathbb{K}v_1$ has the desired property. Suppose therefore that W_{k-1} has been constructed. Then for each associated prime of I_{k-1} of rank $k-1$ there exists some coprime element in V. Taking an appropriate linear combination we find an element v_k in V and not in W_{k-1} which is coprime to all these prime ideals. Hence the ideal I_k defined by (I_{k-1}, v_k) is generated by $W_k = W_{k-1} + \mathbb{K}v_k$ and its rank is k.

For the proof of (iv) we may assume that $\mathfrak{M} = (X_0, X_1-1, \ldots, X_n-1)$ is the ideal of the neutral element of the group G. Then

$$(\mathfrak{M}^c)^h = (X_0, X_1-Y_1, \ldots, X_n-Y_n)$$

and $R/(\mathfrak{M}^c)^h = \mathbb{K}[Y_0, Y_1, \ldots, Y_n]$ is a polynomial ring in $n+1$ variables. Recalling the notation of Section 5.1, it is now obvious that the Poincaré series $P(R/(\mathfrak{M}^c)^h, t)$ is $(1-t^{\varepsilon_0})^{-1}\cdots(1-t^{\varepsilon_n})^{-1}$ and this gives (iv). □

5.4 Calculation of the Jacobi rank

In Section 5.2 we introduced the Jacobi rank ϱ and we gave a lower bound involving ϱ for the length of a certain ideal. We need now to give a precise value for ϱ and we employ in the calculation the concept of the stabiliser of a subvariety of an algebraic group; this reduces the question to the particular case when the ideal belongs to an algebraic subgroup.

Let again G be the algebraic group of Section 5.2. For any subset A of G, the stabiliser of A is defined as the set

$$G_A = \{g \in G;\ g + A \subseteq A\}$$

and it is an algebraic subgroup of G. Further, for an ideal I in the algebra $\mathbb{K}[G]$ of Section 5.3 we denote by $\mathcal{V}(I)$ the variety associated with I and we put $G_I = G_{\mathcal{V}(I)}$. Now writing $W = \mathcal{V}(I)$ and $V = \mathcal{V}(\mathcal{P})$, where $\mathcal{P}$ is a prime component of I with the same rank as I, we call

$$G_{V,W} = \{g \in G;\ g + V \subseteq W\}$$

the transporter of V into W and we observe that it is an algebraic subvariety satisfying $G_V \subseteq G_{V,W}$.

Proposition 5.4 *$G_{V,W}$ is a finite union of cosets of G_V.*

Proof. Let $V_1, \ldots, V_N$ be the irreducible components of W of rank r and let $V = V_1$. Then obviously

$$G_{V,W} = \bigcup_{j=1}^{N} G_{V,V_j}.$$

Now if $g, g' \in G_{V,V_j}$ then we have $g + V \subseteq V_j$ and $g' + V \subseteq V_j$. Hence $g+V = V_j = g'+V$; so $g-g'+V = V$ and therefore $g-g' \in G_V$. Thus $G_{V,V_j} = g + G_V$ for some $g \in G_{V,V_j}$ and this completes the proof. □

We recall that in Section 4.4 we defined the translation operator T_g for any $g \in G$ and the associated comorphism T_g^* of T_g. With this notation we prove the following proposition.

Proposition 5.5 *For any ideal I and associated prime component $\mathcal{P}$ with the rank same as that of I, there exist elements $v_1, \ldots, v_N \in V$ such that the ideal*

$$J = (T_{-v_1}^* I, \ldots, T_{-v_N}^* I)$$

satisfies $\mathcal{V}(J) = G_{V,W}$, where $W = \mathcal{V}(I)$ and $V = \mathcal{V}(\mathcal{P})$.

Proof. Since $G_{V,W} + V \subseteq W$ we have

$$G_{V,W} \subseteq \bigcap_{v \in V} (W - v).$$

On the other hand, if x is an element of the intersection on the right we have $x + v \in W$ for all $v \in V$, whence $x + V \subseteq W$. This means that $x \in G_{V,W}$. Since the Zariski topology is Noetherian, it follows that, for some $v_1, \ldots, v_N \in V$,

$$G_{V,W} = \bigcap_{v \in V} (W - v) = \bigcap_{j=1}^{N} (W - v_j).$$

It is now clear that the ideal J defined in the proposition has the desired properties. □

In the sequel, to simplify notation, we shall extend the functions δ and ϱ to Zariski-closed algebraic subsets V of G by putting $\delta_j(V) = \delta_j(\mathcal{I}(V))$ and $\varrho(V) = \varrho(\mathcal{I}(V))$; here $\mathcal{I}(V)$ is the ideal of elements of $\mathbb{K}[G]$ vanishing on V. Then $\delta_j(V) = 0$ unless $j(0) + \cdots + j(n)$ is equal to the codimension of V. We note that, in this situation, if $G = \mathbb{G}_a \times \mathbb{G}_m^n$ then the polynomial $\chi_{M,d(M)}(\gamma)$ introduced in Section 5.1 can be rewritten as

$$\sum \delta_j \gamma^i / i!,$$

where the sum is taken over all i, j with $i + j = 1$ and $j(0) + \cdots + j(n) = d(M)$.

We have seen in Section 4.4 that the Lie algebra of G is

$$\operatorname{Lie} G = \mathbb{K}\frac{\partial}{\partial X_0} + \mathbb{K}X_1\frac{\partial}{\partial X_1} + \cdots + \mathbb{K}X_n\frac{\partial}{\partial X_n}.$$

We now denote by $\mathfrak{h}$ some subspace of dimension n of the Lie algebra of G and fix a basis $\Delta_0, \ldots, \Delta_{n-1}$ for $\mathfrak{h}$. Then with the Jacobi rank ϱ defined with respect to this subspace we prove the following.

Lemma 5.6 *Let $H \subseteq G$ be a connected algebraic subgroup of G. Then*

$$\varrho(H) = \dim(\mathfrak{h}/(\mathfrak{h} \cap \operatorname{Lie} H)).$$

Further, if $\varrho(H) < \dim(G/H)$ *then* $\operatorname{Lie} H \subseteq \mathfrak{h}$ *and* $\varrho(H) = \dim(\mathfrak{h}/\operatorname{Lie} H)$.

Proof. Let $\chi_1, \ldots, \chi_n$ be a basis for the multiplicative character group $X(\mathcal{T})$ of $\mathcal{T} = \mathbb{G}_m^n$ as indicated in the proof of Lemma 4.4. Further, let χ_0 be a generator for the additive character group $\operatorname{Hom}(\mathbb{G}_a, \mathbb{G}_a)$ so that we can identify X_i with χ_i for $i = 0, \ldots, n$. The subgroup H can be expressed as $H_a \times H_m$ where H_m is a subtorus of $\mathcal{T}$ and $H_a = \mathbb{G}_a$ or $H_a = 0$.

We distinguish two cases. In the first case we have $H_a = \mathbb{G}_a$ and then the ideal $I(H)$ has a basis $\mathcal{B}$ of the form $\chi - 1$ for $\chi \in X(\mathcal{T})$. Further, the embedding $\operatorname{Lie} H \subseteq \operatorname{Lie} G$ is defined by the equations $l_\chi = 0$ for χ an element in the basis and with $l_\chi \in \mathfrak{g}^*$ where $\mathfrak{g} = \operatorname{Lie} G$ (see Section 4.5). Since $\Delta(\chi - 1) = \Delta\chi = l_\chi(\Delta) \cdot \chi$ for $\Delta \in \{\Delta_0, \ldots, \Delta_{n-1}\}$ we have $J(I(H), \mathfrak{h}) = (\Delta\chi)_{\chi,\Delta} = (l_\chi(\Delta))_{\chi,\Delta} \circ U$ where χ runs through the elements of $\mathcal{B}$, Δ runs through the elements $\Delta_0, \ldots, \Delta_{n-1}$ and U is diagonal with $\chi \in \mathcal{B}$ as diagonal elements. This gives

$$\varrho(H) = \operatorname{rank}(l_\chi(\Delta))_{\chi \in \mathcal{B},\ \Delta \in \{\Delta_0, \ldots, \Delta_{n-1}\}}.$$

The latter is equal to the dimension of the image of $\mathfrak{h}$ under the linear map $L = (\ldots, l_\chi, \ldots)_{\chi \in B}$ from $\mathfrak{h}$ to $\mathbb{K}^{n+1-\dim H}$. Since the image of L is isomorphic to $\mathfrak{h}/\ker L = \mathfrak{h}/(\mathfrak{h} \cap \operatorname{Lie} H)$, the required value for $\varrho(H)$ follows. In the second case we have $H_a = 0$. Then one of the basis elements for the ideal of H can be taken as $\chi = \chi_0$ and χ_0 has to be replaced by 1 in the matrix U. We find that

$$\varrho(H) = 1 + \operatorname{rank}(l_\chi(\Delta))_{\chi \in \mathcal{B};\ \chi \neq \chi_0,\ \Delta \in \{\Delta_1, \ldots, \Delta_{n-1}\}}$$

and the latter is again equal to $\dim(\mathfrak{h}/(\mathfrak{h} \cap \operatorname{Lie} H))$. This completes the proof of the first part.

For the proof of the second part we note that

$$\mathfrak{h}/(\mathfrak{h} \cap \operatorname{Lie} H) \cong (\mathfrak{h} + \operatorname{Lie} H)/\operatorname{Lie} H.$$

Now, if $\varrho(H) < \dim(G/H)$ we have $\mathfrak{h} + \operatorname{Lie} H \neq \mathfrak{g}$ and this gives $\operatorname{Lie} H \subseteq \mathfrak{h}$ and $\varrho(H) = \dim(\mathfrak{h}/\operatorname{Lie} H)$. □

5.5 The Wüstholz theory

We shall now prove the main theorem on multiplicity estimates that we need in the application to the theory of logarithmic forms. It is a modified version of the basic result of Wüstholz [262]. The latter is now an essential tool in transcendence theory and it is discussed and utilised widely in the literature; see e.g. [253].

We denote by $\mathbb{K}$ an algebraically closed field of characteristic 0 and by $\mathcal{R}$ the polynomial ring over $\mathbb{K}$ in the variables $X_0, \ldots, X_n$. Points in $G(\mathbb{K})$, where $G = \mathbb{G}_a \times \mathbb{G}_m^n$ is the group of Section 4.3, may be identified with the functions ξ from $\{0, \ldots, n\}$ to $\mathbb{K}$ with function values $\xi(j) \in \mathbb{K}^\times$ for $1 \le j \le n$. For any such ξ we denote by Δ_ξ the endomorphism of $\mathcal{R}$ of the shape $X_0 \mapsto X_0 + \xi(0)$ and $X_j \mapsto \xi(j)X_j$ that fixes $\mathbb{K}$; note that this operator is not differential but it is one induced by a translation operator (see Section 5.4). Further, we denote by $\varepsilon \in G(\mathbb{K})$ the neutral element given by $\varepsilon(0) = 0$ and $\varepsilon(j) = 1$ for $j > 0$.

Let $S > 0$ and $T \ge 0$ be integers. We introduce integers $S_1 \ge \cdots \ge S_{n+1} \ge 0$ and integers $T_1 \ge \cdots \ge T_{n+1} \ge 0$ such that

$$S_1 + \cdots + S_{n+1} \le S, \quad T_1 + \cdots + T_{n+1} \le T .$$

Further, we introduce numbers $D_0 \ge 0, \ldots, D_n \ge 0$ and we suppose that the D_i are not all 0. For each integer k with $1 \le k \le n+1$ and each function $j \in \Gamma(k)$, where $\Gamma(k)$ is defined in Section 5.3, we assume that

$$(S_k + 1)\binom{T_k + k - \delta_{k,n+1}}{k - \delta_{k,n+1}} > k!\, D_0^{j(0)} D_1^{j(1)} \cdots D_n^{j(n)}; \qquad (5.5)$$

here $\delta_{i,j}$ is the Kronecker δ-function. Then, recalling the notation of Lemmas 5.3 and 5.6, we prove the following.

Theorem 5.7 *Suppose that $P(X_0, X_1, \ldots, X_n)$ is a polynomial in $\mathcal{R}$, not identically zero, with*

$$\deg_{X_i} P \le D_i \quad (0 \le i \le n).$$

Suppose further that for any non-negative integers $t(0), \ldots, t(n-1)$ satisfying $t(0) + \cdots + t(n-1) \le T$ and any integer s with $0 \le s \le S$ we have

$$\Delta_\alpha^s \Delta_0^{t(0)} \Delta_1^{t(1)} \cdots \Delta_{n-1}^{t(n-1)} P(\varepsilon) = 0, \qquad (5.6)$$

where $\alpha = (1, \alpha_1, \ldots, \alpha_n) \in G(\mathbb{K})$*. Then there exists a connected algebraic subgroup* H *of* G *with codimension* r *with* $1 \le r \le n$ *such that one of the following holds.*

(i) $\mathrm{Lie}\, H \subseteq \mathfrak{h}$ *and, for all* $j \in \Gamma(r)$*, we have*

$$(S_r + 1)\binom{T_r + r - 1}{r - 1}\delta_j(H) \le r!\, D_0^{j(0)} D_1^{j(1)} \cdots D_n^{j(n)}. \qquad (5.7)$$

(ii) *For some integer* s *with* $0 < s \le S_r$ *we have* $s\alpha$ *an element of* $H(\mathbb{K})$ *and (5.7) holds for all* $j \in \Gamma(r)$ *with* $S_r + 1$ *replaced by* s *and* $r - 1$ *replaced by* $\varrho(H)$.

The theorem is designed to deal with the theory of logarithmic forms as introduced in Section 2.4. It is particularly related to the latter part of the analysis; the earlier part rests on the construction of an auxiliary function and the latter part is sometimes referred to as the deconstruction theory. In this context the subspace $\mathfrak{h}$ is generated by

$$\Delta_0 = \frac{\partial}{\partial X_0}, \qquad \Delta_j = \beta_n X_j \frac{\partial}{\partial X_j} - \beta_j X_n \frac{\partial}{\partial X_n} \quad (1 \le j < n),$$

where $\beta_1, \ldots, \beta_n$ are elements of $\mathbb{K}$ not all zero. In particular we have $\mathrm{Lie}\, \mathbb{G}_a \subseteq \mathfrak{h}$ and thus $H = \mathbb{G}_a \times H_m$. We may assume that the subgroup H in Theorem 5.7 has maximal dimension and we distinguish the cases (i) and (ii). When (ii) holds we have $s\alpha \in H$ and if the α_j are not all roots of unity this implies that $\dim H_m > 0$. The group H_m is defined by equations, holding for α, of the form $\chi^l - 1 = 0$ (see the proof of Lemma 4.4). This leads to multiplicative dependence relations for the α_j as described in Section 2.4. In order to deal with (i) the parameters T, $S, D_0, \ldots, D_n$ are chosen so that the conclusion (5.7) gives $\dim H_m > 0$. Since $\mathrm{Lie}\, H_m \subseteq \mathfrak{h}$ one sees that G can be replaced by G/H_m and one can then apply an inductive procedure.

The whole subject of multiplicity estimates has been created as an alternative approach to the argument involving generalised van der Monde determinants (see Lemma 2.10) which becomes difficult to handle in more far-reaching situations than the one described in Chapter 2. Applications of Theorem 5.7 following the above outline will be discussed in detail in Chapter 6. We remark that the most troublesome factor here is $r!$ that appears on the right-hand side of (5.7) and arises as a consequence of the multihomogeneous degree theory; we have no heuristic explanation as to why this factor should occur and indeed it would seem that it should not be present if one counts conditions as usual in transcendence theory and one believes in the philosophy of Dyson type lemmas. The elucidation of this phenomenon would be of much interest.

Proof of Theorem 5.7. We begin by defining, for $k = 1, \dots, n+1$, parameters

$$S^{(k)} = S - \sum_{j=1}^{k-1} S_j, \quad T^{(k)} = T - \sum_{j=1}^{k-1} T_j.$$

Let now P be the polynomial of Theorem 5.7. We define $I_k(P)$ as the ideal in $\mathcal{R}$ generated by the polynomials $\Delta_\alpha^s \Delta(\tau) P$ where s runs through all integers with $0 \le s \le S - S^{(k)}$ and $\tau(0), \dots, \tau(n-1)$ run through all non-negative integers with $\tau(0) + \cdots + \tau(n-1) \le T - T^{(k)}$; here we have written, for brevity,

$$\Delta(\tau) = \Delta_0^{\tau(0)} \cdots \Delta_{n-1}^{\tau(n-1)}.$$

By (5.3) and (5.5), we have then

$$I_k(P) \subseteq \bigcap_{0 \le s \le S^{(k)}} \mathfrak{M}(s\alpha)^{(T^{(k)})}, \tag{5.8}$$

where $\mathfrak{M}(s\alpha)$ is the maximal ideal of $s\alpha$ in $\mathcal{R}$, that is the set of elements of $\mathcal{R}$ which vanish at $s\alpha$; in fact, $\Delta(\tau)P$ vanishes at $s\alpha$ if and only if $\Delta_\alpha^s \Delta(\tau) P$ vanishes at ε and we then appeal to (5.6).

We show first that if rank $I_{n+1}(P) = n+1$ then the hypothesis (5.5) with $k = n+1$ given at the beginning is false. Indeed, all ideals $I_{n+1}(P)$ and $\mathfrak{M}(s\alpha)$ are then of rank $n+1$ whence we can apply (i) of Lemma 5.3 with $I = I_{n+1}(P)$ and J as the intersection on the right of (5.8) with $k = n+1$. Then, using (ii) of Lemma 5.3 with $\mathfrak{P} = \mathfrak{M}(s\alpha)$ and $\mathfrak{P}^{(t)} = \mathfrak{M}(s\alpha)^{(T^{(k)})}$ and the fact that $\varrho(\mathfrak{M}(s\alpha)) = n$, we get from (5.8)

$$\delta_j(I_{n+1}(P)) \ge \sum_{s=0}^{S'} \delta_j(\mathfrak{M}(s\alpha)) \binom{T'+n}{n},$$

where, for simplicity, we have written $S' = S^{(n+1)}$, $T' = T^{(n+1)}$. Now, by (iv) of Lemma 5.3, we have $\delta_j(\mathfrak{M}(s\alpha)) > 0$ if $j(0)+\cdots+j(n) = n+1$ and hence

$$\delta_j(I_{n+1}(P)) \geq (S'+1)\binom{T'+n}{n}. \tag{5.9}$$

On the other hand, by (iii) of Lemma 5.3, we have

$$\delta_j(I_{n+1}(P)) \leq (n+1)!\, D_0 \cdots D_n. \tag{5.10}$$

From our original inequalities involving S and T we see that $S' \geq S_{n+1}$ and $T' \geq T_{n+1}$ whence, by (5.9) and (5.10), we conclude that

$$(n+1)!\, D_0 \cdots D_n \geq (S'+1)\binom{T'+n}{n} \geq (S_{n+1}+1)\binom{T_{n+1}+n}{n}$$

and this contradicts (5.5) with $k = n+1$ as asserted.

We now consider the set of integers k with $1 \leq k \leq n+1$ such that $\operatorname{rank} I_k(P) \geq k$. The set trivially contains $k = 1$ since $I_1(P) = (P)$. Further, from the argument above the set does not contain $k = n+1$. Hence there is a largest integer $k \leq n$ with this property. Then $k+1 > \operatorname{rank} I_{k+1}(P) \geq \operatorname{rank} I_k(P) \geq k$ whence

$$\operatorname{rank} I_k(P) = k = \operatorname{rank} I_{k+1}(P).$$

This means that there exists some isolated associated prime $\mathfrak{P}$ of $I_k(P)$ of rank k which is also an isolated associated prime of $I_{k+1}(P)$. Thus, since $T^{(k+1)} \geq 0$ and $\mathfrak{P} \supseteq I_{k+1}(P)$, it follows that for $0 \leq s \leq S_k$ and for $\tau(0) + \cdots + \tau(n-1) \leq T_k$ we have

$$\Delta_\alpha^s \Delta(\tau) I_k(P) \subseteq \mathfrak{P}. \tag{5.11}$$

As in Section 5.4 we put $V = \mathcal{V}(\mathfrak{P})$ and $W = \mathcal{V}(I_k)$. Then, by Propositions 5.4 and 5.5, the ideal of $G_{V,W}$ contains

$$J = (\Delta_{-v_1} I_k(P), \ldots, \Delta_{-v_N} I_k(P))$$

and J has rank $k' = \operatorname{cod} G_V$. Furthermore we have $k = \operatorname{cod} V$ whence $k' \geq k$. From (5.11) we deduce that, for the above ranges of s and τ, we have

$$\Delta_\alpha^s \Delta(\tau) J \subseteq (\Delta_{-v_1}\mathfrak{P}, \ldots, \Delta_{-v_N}\mathfrak{P})$$

and, since $v_j \in V$ and so $v_j + G_V \subseteq V$, the ideal on the right is contained in the ideal I of the connected component G_V^0 of the neutral element of the stabilizer G_V of V. This gives

$$J \subseteq \bigcap_{0 \le s \le S_k} (\Delta_{-\alpha}^s I)^{(T_k)}, \tag{5.12}$$

where the bracket notation for the superscript T_k is explained in Section 5.2. It will be observed here that $\Delta_{-\alpha}^s I$ is prime and the radical of $(\Delta_{-\alpha}^s I)^{(T_k)}$.

We now distinguish two cases according as the inequality

$$\Delta_{-\alpha}^s I \ne \Delta_{-\alpha}^{s'} I$$

does or does not hold for all s, s' with $0 \le s < s' \le S_k$. In the first case we have

$$\operatorname{rank}(\Delta_{-\alpha}^s I + \Delta_{-\alpha}^{s'} I) > \operatorname{rank} I$$

and, since $S_k \ge S_{k'}$ and $T_k \ge T_{k'}$, we get by (i)–(iii) of Lemma 5.3

$$\delta_j(J) \ge \delta_j(I)(S_{k'} + 1)\binom{T_{k'} + \varrho'}{\varrho'}, \tag{5.13}$$

where $j \in \Gamma(k')$ and $\varrho' = \varrho(G_V^0)$; note that the latter is equal to $\varrho(G_V)$ since the differential operators are translation invariant. On the other hand, by (iii) of Lemma 5.3 together with (5.5) we see that

$$\delta_j(J) \le k'!\, D_0^{j(0)} \cdots D_n^{j(n)} < (S_{k'} + 1)\binom{T_{k'} + k' - \delta_{k',n+1}}{k' - \delta_{k',n+1}}. \tag{5.14}$$

It is easily verified, by the construction of δ_j in the proof of Lemma 5.3, that we have $\delta_j(I) \ne 0$ for some j whence the inequalities (5.13) and (5.14) cannot both hold unless $k' \le n$ and $\varrho(I) < k'$. By Lemma 5.6 this gives $\varrho(G_V) = k' - 1$ and hence Lie $G_V \subseteq \mathfrak{h}$. Then, from (5.13) and (5.14) again, we obtain (i) of Theorem 5.7 with $H = G_V^0$ and $r = k'$.

In the second case we have $\Delta^{s}_{-\alpha}I = \Delta^{s'}_{-\alpha}I$ for some s, s' in the range $0 \le s < s' \le S_k$. Again we take $H = G^0_V$ and $r = k'$ and we obtain $(s' - s)\alpha \in H(\mathbb{K})$. If we choose s, s' with $s' - s$ minimal we get

$$k'!\, D_0^{j(0)} \cdots D_n^{j(n)} \ge \delta_j(J) \ge (s' - s)\binom{T_{k'} + \varrho'}{\varrho'}\delta_j(I)$$

and (ii) of Theorem 5.7 follows with $s' - s$ in place of s. □

5.6 Algebraic subgroups of the torus

In Section 4.6 we saw that algebraic subgroups of a torus are in bijective correspondence with subgroups of the character group of the torus. Thus we transferred the study of the algebraic subgroups to the simpler and more efficient study of submodules of the character group and, since the latter can be identified with the lattice $\mathbb{Z}^n$, we were able to utilise tools from the Geometry of Numbers. In this way, in Section 4.7 we gave estimates for the size of a basis for a submodule in terms of its weighted volume which depended only on the determinants attaching to the basis. We shall show in this section that the determinants can be replaced by the degrees of the algebraic subgroup as introduced in Lemma 5.3 so that the estimates can be expressed in terms only of these degrees. Since, by the bijective correspondence, a basis for the submodule gives immediately a basis for the ideal defining the subgroup, this furnishes an effective determination in terms of the degrees for the basis of the ideal. In computational algebraic geometry the only other way currently known to obtain estimates of this kind is the theory of Gröbner bases, but one would not get an estimate with a quality that is comparable to ours. We mention that, though the details have yet to be worked out, the method clearly applies not only to tori but also more generally to semi-abelian varieties.

To state the main result, let I be the set $\{1, \ldots, n\}$, let $\mathcal{T} = \mathbb{G}_m^n$ as in Section 4.6 and let $\mathcal{S}$ be an algebraic subgroup of $\mathcal{T}$. We denote by $\mathcal{M}$ the subgroup of elements $l \in \mathbb{Z}^I$ as in the proof of Lemma 4.4 for which $\chi^l - 1 = 0$ on $\mathcal{S}$. Then $\mathcal{M}$ is a free abelian subgroup and its rank is equal to the codimension of the subtorus $\mathcal{S}$. Let $\mathbf{m}_1, \ldots, \mathbf{m}_r$ be a basis for $\mathcal{M}$ and let J be a subset of I. The set $\mathbf{m}_1, \ldots, \mathbf{m}_r$ together with the set, for $\nu \in I \setminus J$, of all ε_ν generate a free subgroup $\mathcal{M}_J$ of $\mathbb{Z}^I$. We define $\mu_J(\mathcal{S})$ to be the index $[\mathbb{Z}^I : \mathcal{M}_J]$ of $\mathcal{M}_J$ in $\mathbb{Z}^I$ if $|J| = r$ and if the rank of $\mathcal{M}_J$ is equal to n; we define it to be 0 in all other cases. The integer $\mu_J(\mathcal{S})$ depends only on $\mathcal{S}$ and J and it is independent of the choice of the basis for $\mathcal{M}$. We also see that $\mu_J(\mathcal{S})$ is equal to the absolute value of the determinant of the $r \times r$ minor $M_{j_1,\ldots,j_r}$ corresponding to the set $J = \{j_1, \ldots, j_r\}$ introduced in Section 4.7.

Now let H be an algebraic subgroup of G whence by Proposition 4.3 there is a decomposition $H = H_a \times H_m$ with $H_a \subseteq \mathbb{G}_a$ and with $H_m \subseteq \mathcal{T}$. Further let δ_j, for $j \in \Gamma(k)$, be the functions given in Lemma 5.3 and denote by J the set of $\nu \in I$ for which $j(\nu) = 1$. We define $\mu_j(H) = \mu_J(H_m)$ for $k = \operatorname{codim} H$ and $\mu_j(H) = 0$ for $k \neq \operatorname{codim} H$. Then the following holds.

Theorem 5.8 *We have $\mu_j(H) \leq \delta_j(H)$.*

Proof. We deal first with the case when H is a finite subgroup. Then $\mu_j(H) = 0$ unless $j = 1$. If $j = 1$, Lemma 5.3 together with (5.1) shows that $\delta_1(H)$ is equal to the dimension of the vector space $\mathbb{K}[G]/\mathcal{I}(H)$ over $\mathbb{K}$. The restrictions, for l in $\mathbb{Z}^I$, of $\chi^l - 1$ to H contain a basis for this vector space and it follows that its dimension is given by $[\mathbb{Z}^I : M(H_m)] = \mu_J(H_m)$. Since $\mu_J(H_m) = \mu_1(H)$ we see that $\delta_1(H) = \mu_1(H)$ and this proves the theorem for finite H.

The general case is proved by induction on the dimension of H. Let the dimension be h and assume that the lemma is established for subgroups with dimension less than h. There are two cases to consider: either $H = 0 \times H_m$ or $H = \mathbb{G}_a \times H_m$. In the first case we have $j(0) = 1$ and we write $j' = j + \varepsilon_\nu$ for some $\nu \geq 1$ such that the ideals $(\chi_\nu - 1)$ and $\mathcal{I}(H_m)$ are coprime. They generate the ideal of some subtorus $H_m' \subseteq H_m$ of dimension $h - 1$ and we deduce from Corollary 5.2 that

$$\delta_{j'}(0 \times H_m') \leq \delta_j(0 \times H_m)\delta_{\varepsilon_\nu}((\chi_\nu - 1)) = \delta_j(0 \times H_m).$$

This inequality is obtained in the same way as indicated in the proof of (iii) of Lemma 5.3 by comparing coefficients in the Poincaré series. The inductive assumption now gives $\delta_{j'}(0 \times H_m') \geq \mu_{j'}(0 \times H_m')$ and the latter is equal to $\mu_j(0 \times H_m)$. In the second case we have $j(0) = 0$ and we see as above that

$$\delta_{j+\varepsilon_0}(0 \times H_m) \leq \delta_j(\mathbb{G}_a \times H_m)\delta_{\varepsilon_0}(0 \times \mathcal{T}) = \delta_j(\mathbb{G}_a \times H_m).$$

From the definition of μ_j we deduce that $\mu_{j+\varepsilon_0}(0 \times H_m) = \mu_j(\mathbb{G}_a \times H_m)$ for all j with $j(0) = 0$ and this completes the proof. □

Finally we discuss the connection with logarithmic forms. We recall that in Section 4.7 we introduced the weighted volume $\mathrm{vol}_{\mathbf{a}}(\mathcal{M})$ of a lattice $\mathcal{M}$ and in Lemma 4.7 we established the existence of a basis for $\mathcal{M}$ such that the product of the norms of the basis elements with respect to a weighted distance function $G_{\mathbf{a}}$ are bounded above in terms of $\mathrm{vol}_{\mathbf{a}}(\mathcal{M})$. In view of Theorem 5.8, the result can now be applied to the lattice $\mathcal{M} = M(H_m)$ consisting of all functions l for which $\chi^l - 1$ vanishes on H_m. This shows that $\mathrm{vol}_{\mathbf{a}}(M(H_m))$ can be estimated in terms of the quantities $\delta_j(H)$ and Theorem 5.7 gives upper bounds for the latter which depend only on the degree of the polynomial P. In the context of the theory of logarithmic forms, the argument is used in the inductive procedure that is referred to in Section 5.5 and which is discussed more fully in Chapter 7.

6
The analytic subgroup theorem

6.1 Introduction

In this chapter we give a detailed exposition of the analytic subgroup theorem which is one of the most significant results in modern transcendence theory. A basic tool in the proof is the theory of multiplicity estimates on group varieties and in the preceding chapter we gave in full detail the simplest version of the theory. The latter is in fact sufficient to establish the results on logarithmic forms to be discussed in Section 7.2. There the underlying algebraic group is of the type $\mathbb{G}_a \times \mathcal{T}$ where $\mathcal{T}$ is a split torus of the form $\mathcal{T} = \mathbb{G}_m^n$ for some integer n. If one wishes to deal with a general commutative group variety one has to expand considerably the results on multiplicity estimates we have obtained so far. This will be presented in a general context in Section 6.7 and we shall postpone further discussions until then.

We begin our exposition by recalling some of the basic facts about group varieties over a field $\mathbb{K}$ of characteristic zero. A group variety is a quasi-projective variety G together with morphisms $+\colon G \times G \to G$ (addition) and $-\colon G \to G$ (inverse) and a neutral element which satisfy the usual group axioms. Since G is a quasi-projective variety there exists a projective space $\mathbb{P}^N$ such that G is a Zariski open subset of the set of zeros $\overline{G}$ of a finite collection of homogeneous polynomials in the variables $X_0, \ldots, X_N$. The variety $\overline{G}$ is called the Zariski closure of G. Further, G being Zariski open in $\overline{G}$ means that the complement of G in $\overline{G}$ is itself the set of zeros of another finite set of homogeneous polynomials.

We may choose the projective space $\mathbb{P}^n$ in such a way that the morphism $+\colon G \times G \to G$ is given by a finite collection of sets of bihomogeneous polynomials $P_0, \ldots, P_N$ in the homogeneous variables $X_0, \ldots, X_N$ and $Y_0, \ldots, Y_N$. This collection has the property that if g, h are points on G with homogeneous coordinates $(g_0, \ldots, g_N)$ and $(h_0, \ldots, h_N)$ then there is a set $P_0, \ldots, P_N$ in the collection such that the homogeneous coordinates of $g + h$ are given by the polynomials evaluated at the point (g, h). In other words

$$(P_0(g_0, \ldots, g_N; h_0, \ldots, h_N), \ldots, P_N(g_0, \ldots, g_N; h_0, \ldots, h_N))$$

are homogeneous coordinates of $g+h$. We say that such a set of polynomials $P_0, \ldots, P_N$ defines an addition formula for $g+h$ so that, since G is quasi-projective over a field, addition on G is given by a finite collection of such addition formulae. Two addition formulae agree at a point (g, h) whenever they are both defined. The inverse morphism $-\colon G \to G$ can be described by formulae in a similar way.

We have already met examples of group varieties in Chapter 4. There we studied in detail the group varieties $\mathbb{G}_a$ and $\mathbb{G}_m$ which are both contained in the projective space $\mathbb{P}^1$. Their Zariski closure is the whole space $\mathbb{P}^1$. The complement of $\mathbb{G}_a$ and $\mathbb{G}_m$ in $\mathbb{P}^1$ is given by the equations $X_0 = 0$ and $X_0 \cdot X_1 = 0$ respectively; taking products gives further examples. Finite products of the additive group are called vector groups and finite products of the multiplicative groups are called tori. Other examples are elliptic curves and, more generally, abelian varieties; they can be given explicitly as the sets of zeros of a finite number of homogeneous polynomials. Their theory was studied in the nineteenth century by Weierstrass, Riemann and others and, in particular, addition formulae were known at that time. However, only in the past decade have complete sets of addition formulae been constructed explicitly. This goes back to the work of Igusa and Mumford and we refer to Section 6.9 for further details.

The commutative group varieties that have been introduced so far form the building blocks of a general commutative group variety. We know from a classical result of Rosenlicht that an arbitrary commutative group variety G is an extension of an abelian variety A by a linear algebraic group L which is a product of a vector group V and a torus $\mathcal{T}$.

We obtain therefore an exact sequence

$$0 \longrightarrow V \times \mathcal{T} \longrightarrow G \longrightarrow A \longrightarrow 0$$

of commutative algebraic groups. Here the arrows are morphisms which have a description by homogeneous polynomials similar to the morphisms appearing in the definition of a group variety.

The Lie algebra of a group variety is defined as the set of translation invariant vector fields on this variety. In Chapter 4 we determined the Lie algebra in the cases when the group variety is additive and multiplicative. It can also be determined explicitly for abelian varieties; however, this becomes more difficult since the addition formulae are more complicated and addition cannot be described by just one formula. Fortunately there is no need for us to discuss the definition further here since the only significant property of the Lie algebra that we shall use is that it is a vector space over the field of definition of the group variety.

The set of complex points $G(\mathbb{C})$ on a group variety G for a given embedding of $\mathbb{K}$ into $\mathbb{C}$ has the structure of a complex manifold. Further, on noting that the addition and inverse maps are, in this case, holomorphic, it follows that we can regard the group variety as a complex Lie group. The Lie algebra of the latter is defined analogously to the Lie algebra of a group variety and again we do not discuss this further since we shall need only the property that it is a finite dimensional complex vector space. If we denote by $\mathfrak{g}$ the Lie algebra of the group variety G then the Lie algebra of the complex Lie group associated with G is the complex vector space $\mathfrak{g} \otimes \mathbb{C}$; the tensor product here is taken over the field of definition of the group G. Both the Lie group and its Lie algebra are related by the exponential map $\exp_G\colon\ \mathfrak{g} \otimes \mathbb{C} \rightarrow G(\mathbb{C})$. This is a holomorphic homomorphism from the Lie algebra into the group and is defined using one-parameter subgroups. Namely, given any element ξ in the Lie algebra there exists by the theory of linear differential equations a unique homomorphism

$$\varphi_\xi : \mathbb{C} \longrightarrow G(\mathbb{C})$$

such that its differential has the property that it maps the vector field dt to ξ. The exponential map is then the unique map taking the vector field

ξ to $\varphi_\xi(1)$. It is established as a fundamental result of Lie group theory that this leads to an analytic homomorphism $\exp_G$.

As before, let $X_0, \ldots, X_N$ be projective coordinates for the projective space containing the group variety G. Then the holomorphic functions

$$f_i = \exp_G^*(X_i) = X_i \circ \exp_G$$

can be determined explicitly; this is described in full detail in [92]. Particular examples of these f_i are the linear function z when $G = \mathbb{G}_a$, the exponential function e^z when $G = \mathbb{G}_m$ and the theta-functions when G is an abelian variety. In general the f_i are built up from these functions and, as a consequence, the order of growth is at most 2. This makes it clear that one can study transcendence properties of their values and in fact many questions of this type have been considered in the past. We mention only the transcendence of e and π, of α^β and the well-known discoveries of Siegel [229] and Schneider [213] on the periods of elliptic integrals of the first and second kind. All these results can be deduced from the analytic subgroup theorem which we now discuss.

Let G be a commutative algebraic group defined over a number field $\mathbb{K}$ with Lie algebra $\mathfrak{g}$. Further, let $\mathfrak{b}$ be a subalgebra of $\mathfrak{g}$ and put $B = \exp_G(\mathfrak{b} \otimes_{\mathbb{K}} \mathbb{C})$. Then B is a Lie subgroup of $G(\mathbb{C})$ but not necessarily closed. We call B an analytic subgroup of G. Since $\mathfrak{b}$ is a vector space over $\mathbb{K}$ we say that the analytic subgroup B is defined over $\mathbb{K}$. We want to determine the group of algebraic points

$$B(\overline{\mathbb{K}}) = B \cap G(\overline{\mathbb{K}})$$

on B. One observes at once that this group is non-trivial provided that there exists a non-trivial algebraic subgroup H of G defined over a number field such that $H(\mathbb{C}) \subseteq B$. In fact we have then

$$H(\overline{\mathbb{K}}) \subseteq B(\overline{\mathbb{K}}) \quad \text{and} \quad H(\overline{\mathbb{K}}) \neq 0.$$

The following gives the converse statement.

Theorem 6.1 (Analytic subgroup theorem) *Let $B \subseteq G(\mathbb{C})$ be an analytic subgroup of $G(\mathbb{C})$ defined over $\mathbb{K}$. Then $B(\overline{\mathbb{K}}) \neq 0$ if and only if there exists a non-trivial algebraic subgroup $H \subseteq G$ defined over a number field such that $H(\mathbb{C}) \subseteq B$.*

As already indicated, it is the ‘only if’ implication here that is non-trivial. The analytic subgroup theorem is the most natural generalisation in terms of algebraic groups of the qualitative version of Baker’s theorem on linear forms in logarithms; we shall give a detailed outline of the proof in Section 6.8 where, as we shall see, a major role is played by multiplicity estimates on group varieties. Incidentally, the first published reference to algebraic groups in the context of transcendence theory seems to be due to Lang [134]; he himself attributes the original thought to a problem of Cartier concerning the Lindemann theorem and he gives a discussion of the latter and also the Gelfond–Schneider theorem in terms of group varieties.

We shall now discuss some of the most important consequences of Theorem 6.1. To begin with, the classical results on the transcendence of special values of the ordinary exponential function due to Hermite, Lindemann, Gelfond and Schneider are, as mentioned above, all contained in the theorem. For example we know from Lindemann that if α is any non-zero complex number then not both of α and e^α can be algebraic. This can be deduced from the analytic subgroup theorem as follows. Assume the contrary, namely that both α and e^α are algebraic, and consider the algebraic group $G = \mathbb{G}_a \times \mathbb{G}_m$ which is clearly defined over the rational numbers. From Section 4.4 we see that its Lie algebra is given up to isomorphism by

$$\mathfrak{g} = \mathbb{Q} \times \mathbb{Q}$$

and if we denote as usual by $\mathbb{C}^*$ the multiplicative group of complex numbers we have

$$G(\mathbb{C}) = \mathbb{C} \times \mathbb{C}^* .$$

The exponential map $\exp_G$ is given by

$$\exp_G \colon \mathbb{C} \times \mathbb{C} \longrightarrow \mathbb{C} \times \mathbb{C}^*, \quad (z, w) \longmapsto (z, e^w)$$

and clearly $\mathbb{C} \times \mathbb{C} = \mathfrak{g} \otimes_{\mathbb{Q}} \mathbb{C}$. Let $\Delta \subseteq \mathbb{C} \times \mathbb{C}$ be the diagonal and let $B = \exp_G(\Delta)$ be the associated analytic subgroup. Then B is connected and has dimension 1. Further, one has $(\alpha, \alpha) \in \Delta$ and $(\alpha, e^\alpha) = \exp_G(\alpha, \alpha) \in B(\overline{\mathbb{Q}})$; since $(\alpha, e^\alpha) \neq (0, 1)$, this means that $B(\overline{\mathbb{Q}})$ is non-trivial. Hence there exists a proper algebraic subgroup H

of G such that $H(\mathbb{C}) \subseteq B$. The Lie group $H(\mathbb{C})$ is non-trivial whence it has dimension at least 1 and, since it is a subgroup of B, it must have dimension at most 1. It follows that $H(\mathbb{C}) = B$, in other words B is an algebraic subgroup. Since B is the graph of the exponential function e^z this implies that e^z is an algebraic function which is evidently not the case. Therefore not both of α and e^α can be algebraic. From this one deduces the transcendence of e and π as well as that of e^α for algebraic $\alpha \neq 0$.

We now show how the analytic subgroup theorem implies the qualitative version of Baker's theorem on logarithmic forms. Accordingly let $\alpha_1, \ldots, \alpha_n$ be algebraic numbers, not 0 or 1, and let $L = L(z_1, \ldots, z_n) \neq 0$ be a linear form in the variables $z_1, \ldots, z_n$ with coefficients in an algebraic number field $\mathbb{K}$ which we assume contains $\alpha_1, \ldots, \alpha_n$. Then if

$$L(\log \alpha_1, \ldots, \log \alpha_n) = 0$$

Baker's theorem tells us that there exists a linear form $M(z_1, \ldots, z_n) \neq 0$ with integer coefficients such that

$$M(\log \alpha_1, \ldots, \log \alpha_n) = 0.$$

Let now G be the algebraic torus $\mathbb{G}_m^n$ with character group $X(G)$ and let $\mathfrak{g}$ be its Lie algebra which is clearly equal to $\mathbb{Q}^m$ up to isomorphism. Further, let $\exp_G$ be the exponential map

$$(z_1, \ldots, z_n) \in \mathbb{C}^n \mapsto (\exp(z_1), \ldots, \exp(z_n)) \in \left(\mathbb{C}^*\right)^n$$

and note that $\mathbb{C}^n = \mathfrak{g} \otimes_{\mathbb{Q}} \mathbb{C}$. Then the equation $L = 0$ defines a subspace $\mathfrak{b}$ of $\mathfrak{g} \otimes_{\mathbb{Q}} \mathbb{K}$ and thus also an analytic subgroup $B = \exp_G(\mathfrak{b} \otimes_{\mathbb{K}} \mathbb{C})$ of $G(\mathbb{C}) = (\mathbb{C}^*)^n$. Since the vector $u = (\log \alpha_1, \ldots, \log \alpha_n)$ is in $\mathfrak{b} \otimes_{\mathbb{K}} \mathbb{C}$ and $\exp(u) = (\alpha_1, \ldots, \alpha_n)$ is in $B(\overline{\mathbb{K}})$ and not equal to $(1, \ldots, 1)$, it follows that there exists a non-trivial connected algebraic subgroup H of G with Lie algebra $\mathfrak{h} \subseteq \mathfrak{b}$ such that $H(\mathbb{C}) \subseteq B$. Now the subgroup H of $G = \mathbb{G}_m^n$ is a subtorus of codimension $h < n$ and therefore by Lemma 4.4 there exist linear forms $L_1, \ldots, L_h$ in $z_1, \ldots, z_n$ with integer coefficients such that $\mathfrak{h}$ is given by $L_1 = \cdots = L_h = 0$. Since $\mathfrak{h}$ is contained in $\mathfrak{b}$ we find that L is in the vector space generated by $L_1, \ldots, L_h$ over $\mathbb{K}$. This means that

$$L = \beta_1 L_1 + \cdots + \beta_h L_h$$

with $\beta_1, \ldots, \beta_h \in \mathbb{K}$. If we replace $\alpha_1, \ldots, \alpha_n$ by $\alpha'_1, \ldots, \alpha'_h$ where

$$\alpha'_j = \exp\left(L_j(\log\alpha_1, \ldots, \log\alpha_n)\right)$$

then the new linear form

$$L' = \beta_1 z'_1 + \cdots + \beta_h z'_h$$

has the same properties with respect to $\alpha'_1, \ldots, \alpha'_h$ as L has with respect to $\alpha_1, \ldots, \alpha_n$. By induction, after at most $n-1$ steps, we obtain a non-zero linear form $L'' = \beta z$ and an algebraic number $\alpha'' = \alpha_1^{k_1} \cdots \alpha_n^{k_n}$ in the multiplicative subgroup of $\mathbb{K}^*$ generated by $\alpha_1, \ldots, \alpha_n$. It satisfies

$$L''(\log\alpha'') = 0$$

and hence $\log\alpha'' = 0$; this can be rewritten as

$$k_1 \log\alpha_1 + \cdots + k_n \log\alpha_n = 0$$

for integers $k_1, \ldots, k_n$ not all zero and this is the dependence relation over the integers that we were seeking.

All the classical results on periods of elliptic functions are contained as special cases of the analytic subgroup theorem. As an example we deduce the transcendence of the non-zero periods of the Weierstrass $\wp$-function obtained by Siegel in the case when there is complex multiplication and by Schneider in the general case. We consider the $\wp$-function associated with a lattice Λ in $\mathbb{C}$. Let ω_1 and ω_2 be generators for Λ with $\mathrm{Im}(\omega_2/\omega_1) > 0$ so that $\Lambda = \mathbb{Z}\omega_1 + \mathbb{Z}\omega_2$. Let

$$g_2 = g_2(\Lambda) = \frac{1}{60}\sum \omega^{-4}, \quad g_3 = g_3(\Lambda) = \frac{1}{140}\sum \omega^{-6}$$

be the standard Eisenstein series where the sums are over all non-zero elements $\omega \in \Lambda$. We assume that g_2 and g_3 are contained in an algebraic number field $\mathbb{K}$. Let $\wp(z) = \wp(z, \Lambda)$ be the Weierstrass elliptic function given by

$$\wp(z) = \frac{1}{z^2} + \sum_{0 \neq \omega \in \Lambda} \left(\frac{1}{(z-\omega)^2} - \frac{1}{\omega^2}\right).$$

This is periodic with periods $\omega \in \Lambda$ and it satisfies an algebraic differential equation

$$(\wp'(z))^2 = 4(\wp(z))^3 - g_2\wp(z) - g_3 .$$

Thus, since

$$g_2^3 - 27g_3^2 \neq 0,$$

it follows that the functions $\wp(z)$, $\wp'(z)$ parametrize the elliptic curve E defined over $\mathbb{K}$ given by

$$y^2 = 4x^3 - g_2x - g_3 .$$

It is well known that E is a commutative algebraic group of dimension 1. This follows either from the geometric secant construction or from the addition law for the Weierstrass $\wp$-function. The complex Lie algebra for E is canonically isomorphic to $\mathbb{C}$ and the exponential map $\exp_E$ is given by

$$\exp_E\colon \mathbb{C} \to E(\mathbb{C}) \subseteq \mathbb{P}^2(\mathbb{C}), \quad z \mapsto [\wp(z), \wp'(z), 1].$$

We can now apply the analytic subgroup theorem to prove that any non-zero period ω is transcendental. In fact let $G = \mathbb{G}_a \times E$ so that $G(\mathbb{C}) = \mathbb{C} \times E(\mathbb{C})$. The Lie algebra of the complex Lie group $G(\mathbb{C})$ is given by $\mathbb{C} \times \mathbb{C}$. Again we take Δ as the diagonal in $\mathbb{C} \times \mathbb{C}$ which defines an analytic subgroup B of $G(\mathbb{C})$ of dimension 1. If we assume that ω is algebraic, non-zero and contained in $\mathbb{K}$ we find that, on writing $\gamma = \frac{1}{2}\omega$, the point

$$\exp_G(\gamma, \gamma) = \big(\gamma, [\wp(\gamma), \wp'(\gamma), 1]\big)$$

is in $B(\overline{\mathbb{K}})$. It is known that $\wp(\gamma)$ is a root of the polynomial $4x^3 - g_2x - g_3$ whence $\wp'(\gamma) = 0$ and so all the coordinates on the right-hand side are algebraic numbers. The hypothesis of the analytic subgroup theorem is therefore fulfilled and we obtain a non-trivial algebraic subgroup H of G such that $H(\mathbb{C}) \subseteq B$. Again we conclude that $H(\mathbb{C}) = B$ so that B is algebraic. Hence, as in the first example, it follows that the function $\wp(z)$ is algebraic. This is a contradiction since $\wp(z)$ has infinitely many poles and so we conclude that ω is transcendental.

6.2 New applications

In this section we apply the analytic subgroup theorem to obtain some new results which were not accessible by classical methods in their full generality. The first application deals with elliptic analogues of Baker's theorem in qualitative form which, as we recall, states that if $\alpha_1, \ldots, \alpha_n$ are non-zero algebraic numbers then their logarithms $\log \alpha_1, \ldots, \log \alpha_n$ are linearly independent over the field of algebraic numbers if and only if they are linearly independent over the field of rational numbers.

Some initial steps towards the derivation of elliptic analogues were taken by Baker [20]; here it was shown that the basic analytic techniques could be extended from the classical exponential function to the Weierstrass functions and it was proved that any non-zero expression $\alpha\omega_1 + \beta\omega_2$ is transcendental, where α, β are algebraic and ω_1, ω_2 are periods of two elliptic curves, possibly identical, defined over the algebraic numbers. The subject was substantially developed by Masser in his tract [162]; he proved the analogue of Baker's theorem for elliptic logarithms defined with respect to an elliptic curve with complex multiplication. The theory of multiplicity estimates was not yet available at the time but the hypothesis of complex multiplication made it possible to utilise ad hoc arguments. The work was complemented by Bertrand and Masser [40]; they obtained the desired result for elliptic curves with no complex multiplication whence it follows that the analogue of Baker's theorem holds in general.

We begin by establishing the latter result directly from the analytic subgroup theorem. We fix a number field $\mathbb{K}$ and an elliptic curve E defined over $\mathbb{K}$. This is an algebraic group of dimension 1 and we denote by $\mathfrak{e}$ its Lie algebra. The complex points on the elliptic curve form a complex Lie group $E(\mathbb{C})$ with complex Lie algebra $\mathfrak{e}_{\mathbb{C}} = \mathfrak{e} \otimes_{\mathbb{K}} \mathbb{C}$. The exponential map $\exp_E$ was described in the previous section; its non-trivial components are the Weierstrass $\wp$-function together with its first derivative. We let now $\gamma_1, \ldots, \gamma_n$ be elements of $\mathfrak{e}_{\mathbb{C}}$ such that $\exp_E(\gamma_j)$ is contained in $E(\overline{\mathbb{K}})$ for $j = 1, \ldots, n$ (thus $\gamma_1, \ldots, \gamma_n$ can be interpreted as elliptic logarithms). Further, we denote by $\mathrm{End}(E)$ the ring of endomorphisms of E and we put $\mathbb{K}_0 = \mathrm{End}(E) \otimes \mathbb{Q}$. Then it is known, and in fact easy to prove, that $\mathbb{K}_0$ is either the field $\mathbb{Q}$ or an imaginary quadratic extension of $\mathbb{Q}$; in the latter case we say that the elliptic curve has complex multiplication. The field $\mathbb{K}_0$ has a representation on the Lie algebra $\mathfrak{e}_{\mathbb{C}}$ by scalar multiplication in a natural way.

Theorem 6.2 *The elements $\gamma_1, \ldots, \gamma_n$ of $\mathfrak{e}_{\mathbb{C}}$ are linearly independent over the field of algebraic numbers $\overline{\mathbb{Q}}$ if and only if they are linearly independent over the field $\mathbb{K}_0$.*

Proof. This follows on similar lines to the deduction of Baker's theorem from Theorem 6.1. First we may assume without loss of generality that the vector $\gamma = (\gamma_1, \ldots, \gamma_n)$ is not zero, for otherwise the theorem follows trivially. Suppose now that the numbers in question are linearly dependent over the field $\overline{\mathbb{K}}$. This means that there exists a non-zero linear form $L = L(z_1, \ldots, z_n)$ with coefficients in $\overline{\mathbb{K}}$ such that $L(\gamma_1, \ldots, \gamma_n) = 0$. The linear form defines as usual an analytic subgroup B of the algebraic group $G = E^n$ which contains non-trivial algebraic points, namely all points of the form $\exp_G(r\gamma)$ for $r \in \mathbb{Q}$.

Let now H be the largest algebraic subgroup H of G such that the Lie algebra $\mathfrak{h}$ of H is contained in the Lie algebra $\mathfrak{b}$ of B. We shall apply the analytic subgroup theorem with G replaced by G/H and $\mathfrak{b}$ by $\mathfrak{b}/\mathfrak{h}$. We take $\overline{\gamma}$ as the image of γ in $\mathfrak{b}/\mathfrak{h}$. Clearly the image of $\overline{\gamma}$ under the exponential map of G/H is contained in $(B/H)(\overline{\mathbb{Q}})$ where B/H signifies the analytic subgroup of G/H determined by $\mathfrak{b}/\mathfrak{h}$. The subgroup H was chosen in such a way that no non-trivial algebraic subgroup $\overline{H}$ of G/H has the property that $\overline{H}(\mathbb{C}) \subseteq B/H$. Thus, according to the analytic subgroup theorem, we have $\overline{\gamma} = 0$ whence $\gamma \in \mathfrak{h}$.

The algebraic group H corresponds to an element π of the algebra of endomorphisms $\mathrm{End}(G) \otimes \mathbb{Q}$ of G given by the projection from G to $H \subseteq G$. The algebra of endomorphisms $\mathrm{End}(G) \otimes \mathbb{Q}$ is represented on the Lie algebra of G by the matrix algebra $M_n(\mathbb{K}_0)$ and this means that the endomorphism $\mathrm{id} - \pi$ can be written as an $n \times n$ matrix β with entries in $\mathbb{K}_0$. Since γ is contained in the Lie algebra $\mathfrak{h}$ of H, the element γ is in the kernel of the endomorphism of $\mathfrak{g}$ given by the matrix β. In other words we have found a set of dependence relations

$$\gamma \circ \beta = 0.$$

At least one column of the matrix β is non-zero since H is a proper subgroup of G. We have therefore exhibited a non-trivial dependence relation over $\mathbb{K}_0$ among the elements $\gamma_1, \ldots, \gamma_n$. This proves the theorem. □

The proof given here can easily be extended to cover the case of not just one elliptic curve but more generally a product of the form $G = E_1 \times \cdots \times E_n$. We first group together the pairwise isogenous factors and then notice that the ring of endomorphisms is a product of matrix rings over fields of the form $\mathbb{K}_0$; apart from this the demonstration follows along exactly the same lines.

Our next application concerns the Weierstrass ζ-function $\zeta(z)$ which, we recall, satisfies $\zeta'(z) = -\wp(z)$. Here $\wp(z)$ denotes the Weierstrass $\wp$-function and we assume that the corresponding elliptic curve E is defined over a number field $\mathbb{K}$. The first transcendence results in this context were obtained by Schneider [211, II]; he proved that if ω is a primitive period of $\wp(z)$ and η is the corresponding quasi-period of the associated ζ-function, so that $\zeta(\frac{1}{2}\omega) = \frac{1}{2}\eta$, then any non-vanishing linear combination of ω and η over $\overline{\mathbb{Q}}$ is transcendental. Baker [18] studied the case of two possibly distinct $\wp$-functions as above and succeeded in showing that any non-vanishing linear combination of $\omega_1, \omega_2, \eta_1, \eta_2$ over $\overline{\mathbb{Q}}$ is transcendental; here ω_1, ω_2 denote primitive periods of the respective $\wp$-functions and η_1, η_2 the corresponding quasi-periods. In the special case when ω_1, ω_2 are a pair of fundamental periods of a single $\wp$-function, Coates [71] gave the same result for the five numbers ω_1, ω_2, η_1, η_2 and $2\pi i$, and the definitive theorem in this direction was subsequently proved by Masser [162]; he showed that the dimension of the space spanned by $\omega_1, \omega_2, \eta_1, \eta_2$ together with 1 and $2\pi i$ is either 4 or 6 according to whether $\wp$ does or does not admit complex multiplication.

Now let $\gamma_1, \ldots, \gamma_n$ be elements as above in the complex Lie algebra of E with $\exp_E(\gamma_j) \in E(\overline{\mathbb{K}})$. We shall use the analytic subgroup theorem to determine the dimension of the vector space V generated over $\overline{\mathbb{Q}}$ by the set S of numbers

$$1,\ 2\pi i,\ \gamma_1, \ldots, \gamma_n,\ \zeta(\gamma_1), \ldots, \zeta(\gamma_n)\ ;$$

here we assume that $\gamma_1, \ldots, \gamma_n$ are not in the period lattice of E so that the values of the ζ-function are well-defined. Let r be the dimension of the vector space generated over $\mathbb{K}_0$ by the elements $\gamma_1, \ldots, \gamma_n$. We prove the following theorem.

Theorem 6.3 $\dim V = 2r + 2$.

Proof. The argument follows that of Theorem 6.2 but with a more complicated group variety G. This is given by an extension of the group variety $G_0 = \mathbb{G}_m \times E^n$ by the additive group $\mathbb{G}_a$; see [220]. Its extension class is determined as follows. Let $M(z)$ be a non-zero linear form in the $2n+2$ variables T_{-1}, T_0, $T_1', \ldots, T_n'$, $T_1'', \ldots, T_n''$ with coefficients in $\overline{\mathbb{K}}$ such that M vanishes on the set S. We shall assume that the elements $\gamma_1, \ldots, \gamma_n$ are linearly independent over the field $\mathbb{K}_0$; this involves no loss of generality in view of the functional equation for the ζ-function and the fact that $\exp_E(\gamma_j)$ is in $E(\overline{\mathbb{K}})$. We write M in the form $M = M_0 + M' + M''$ where $M_0 = \alpha T_{-1} + \beta T_0$ and where M', M'' are linear forms in $T_1', \ldots, T_n'$ and $T_1'', \ldots, T_n''$ respectively. The group variety G referred to at the beginning is now given by the extension of G_0 by $\mathbb{G}_a$ determined by M''.

The components of the exponential map $\exp_G$ in the standard embedding (see [92]) are given by the functions

$$z_{-1} + M''(\zeta(z_1), \ldots, \zeta(z_n)),$$
$$\exp(z_0),\ \wp(z_j),\ \wp'(z_j)\ (j = 1, \ldots, n)$$

for complex variables $z_{-1}, z_0, \ldots, z_n$. From the hypotheses it follows that the images of the rational multiples of the point

$$\varepsilon = \big(2\pi i\beta + M'(\gamma_1, \ldots, \gamma_n),\ 2\pi i,\ \gamma_1, \ldots, \gamma_n\big)$$

under the exponential map are algebraic points on G. The rational multiples of ε are all contained in the Lie algebra of an analytic subgroup B which is given by the equation

$$T_{-1} = \beta T_0 + M'.$$

This subalgebra is clearly defined over the field $\overline{\mathbb{K}}$; therefore we can apply the analytic subgroup theorem and deduce that there exists an algebraic subgroup H of G defined over $\overline{\mathbb{K}}$ whose complex Lie algebra $\mathfrak{h}_{\mathbb{C}}$ is contained in the Lie algebra of B. Let $\pi : G \to E^n$ be the canonical homomorphism given by the Rosenlicht exact sequence in Section 6.1 and let $p \colon E^n \to \pi(H) \subseteq E^n$ be the projector in $\mathrm{End}(E^n)$ determined by H. Then H is contained in the kernel of $q = 1 - p$ and hence $q(\pi(g)) = 0$. Thus we obtain $d(q \circ \pi)(\gamma) = 0$ and this is a linear dependence relation over $\mathbb{K}_0$ between $\gamma_1, \ldots, \gamma_n$. Since the latter are linearly independent over $\mathbb{K}_0$ it follows that $d(q \circ \pi) = 0$ and this gives $q = 0$ or equivalently $p = \mathrm{id}$. Hence we have $\pi(H) = E^n$.

We assume now that $\beta \neq 0$. In fact otherwise we can take the quotient of G by the multiplicative group $\mathbb{G}_m$ and we are in a simpler situation which can be dealt with in the same way as the general case. The intersection of H with $\mathbb{G}_a \times \mathbb{G}_m$, that is the kernel of π, has the form $H_a \times H_m$ by Proposition 4.3 and it is contained in the intersection of B with $\mathbb{G}_a \times \mathbb{G}_m$. Thus, as in the deduction of Lindemann's theorem from the analytic subgroup theorem, we see that $H_a \times H_m$ is finite. Hence the canonical projection from G to E^n makes H into a covering of E^n. This means that the projection has a section which is a homomorphism and therefore the extension class of G is zero. It follows that the linear form M'' which defines the extension class is itself zero. Hence $M = M_0 + M'$. However, since H is isogenous to E^n, it is defined in the Lie algebra by the equations $T_{-1} = T_0 = 0$. Since further the Lie algebra of H is contained in that of B, the linear forms M, T_{-1}, T_0 are linearly dependent. This implies that $M' = 0$ identically whence $M = M_0$. But the number π is transcendental and therefore we must have $\beta = 0$ in contradiction to our assumption. This proves the theorem. □

So far we have dealt only with the case of one elliptic curve. However, there is no obstruction to obtaining an analogous result in the general case. Accordingly let now $E_1, \ldots, E_n$ be elliptic curves over the field $\overline{\mathbb{K}}$ and γ_j be an element in the Lie algebra of E_j $(j = 1, \ldots, n)$ with the property that $\exp_{E_j}(\gamma_j) \in E_j(\overline{\mathbb{K}})$. We let I_ν $(\nu = 1, \ldots, k)$ be maximal sets of indices for which the corresponding elliptic curves are pairwise isogenous and we take $E^{(\nu)}$ to be an elliptic curve in the set $\{E_j; j \in I_\nu\}$ for $\nu = 1, \ldots, k$. Under an arbitrarily chosen isogeny from E_j to $E^{(\nu)}$, the images of the elements γ_j $(j \in I_\nu)$ in the tangent space of $E^{(\nu)}$ generate a vector space Γ_ν; the latter is defined over the field $\mathbb{K}_0^{(\nu)}$ obtained from the ring of endomorphisms of $E^{(\nu)}$. Finally we define ϱ_ν as the dimension of Γ_ν over $\mathbb{K}_0^{(\nu)}$ and we take ϱ as the dimension of the vector space spanned by the set S defined prior to Theorem 6.3. Then the following result holds.

Theorem 6.4 *We have* $\varrho = 2(\varrho_1 + \cdots + \varrho_k) + 2$.

The proof is left as an exercise for the ambitious reader. A result of this kind was first established in Wüstholz [261]; here he extended Baker's result [18] on the periods and quasi-periods of a product of two elliptic curves to the more general case of a product of an arbitrary number of elliptic curves. This solved the 'period problem' originally proposed by Baker.

We proceed now to discuss a further application of the analytic subgroup theorem; it deals with a problem that Schneider raised in his book [216] about transcendence properties of elliptic integrals of the third kind. There was no apparent progress for a long time but, in 1980, Laurent [141] surprisingly succeeded in obtaining the first results in this direction. They needed strong restrictions on the poles of the differential form of the third kind; the latter were removed by Wüstholz [260]. Like [261], this was a major work that utilised Baker's method combined with the then recently discovered theorems concerning multiplicity estimates on group varieties. In order to state the results of [260] we take an elliptic curve E defined over a number field $\mathbb{K}$ given as usual by an equation of the form

$$y^2 = 4x^3 - g_2 x - g_3$$

with $g_2, g_3 \in \mathbb{K}$. We denote by Λ the lattice comprising the periods of the differential form dx/y of the first kind; let $\wp(z)$, $\zeta(z)$ and $\sigma(z)$ be the classical Weierstrass functions associated with Λ. We fix a non-zero period ω in Λ and take $\eta = \eta(\omega)$ to be the corresponding quasi-period of $\zeta(z)$; this satisfies $\eta(\omega) = \zeta(z+\omega) - \zeta(z)$. For any complex number u not in Λ we define

$$\lambda(u, \omega) = \omega\zeta(u) - \eta u.$$

Let now $u_1, \ldots, u_n$ be complex numbers, not in Λ, such that $\wp(u_1), \ldots, \wp(u_n)$ are algebraic. Then we have the following result.

Theorem 6.5 *Any non-vanishing linear form in* ω, η, $\lambda(u_1, \omega), \ldots, \lambda(u_n, \omega)$ *with algebraic coefficients is transcendental.*

As an application, consider any meromorphic differential ξ on the elliptic curve E which is defined over the same field as the curve itself.

Then, by standard manipulations, ξ can be written in the shape

$$\xi = \frac{1}{2}\sum_{j=1}^{n} c_j \frac{y + y_j}{x - x_j}\frac{dx}{y} + a\frac{dx}{y} + bx\frac{dx}{y} + d\chi\,,$$

where χ is a rational function on E and the (x_j, y_j) are points on E. Since the differential form is defined over an algebraic number field, it follows that the c_j, x_j, y_j and a, b are all algebraic. We choose complex numbers u_j such that $\wp(u_j) = x_j$; this can be done in a unique way modulo Λ. If γ is any closed cycle on $E(\mathbb{C})$ along which ξ is holomorphic then the period of the integral of ξ along the cycle takes the form

$$\sum_{j=1}^{n} c_j(\lambda(u_j, \omega) + 2k_j\pi i) + a\omega + b\eta$$

for some integers k_j $(1 \leq j \leq n)$. Except for the term $2\pi i$, the latter is a linear expression in the quantities appearing in Theorem 6.5. To incorporate the $2\pi i$ we appeal to Legendre's relation

$$\eta_1\omega_2 - \eta_2\omega_1 = 2\pi i$$

satisfied by the fundamental periods and quasi-periods of the Weierstrass functions; this gives $\lambda(u', \omega)$ as a rational multiple of $2\pi i$ on taking $u' = \frac{1}{2}\omega'$ where ω' is a primitive period in Λ such that ω, ω' are linearly independent. Thus on applying Theorem 6.5 with the additional quantity $\lambda(u', \omega)$ we obtain the following.

Corollary 6.6 *The periods of ξ are either zero or transcendental.*

We recall that this result was originally proved by Siegel [229] in the special case when ξ is a differential form of the first kind on a curve with complex multiplication. The latter condition was subsequently removed by Schneider [213] who also dealt with the case of differential forms of the second kind. As mentioned earlier, the first non-trivial results on periods of elliptic integrals of the third kind were given by M. Laurent [141]; his work involved certain restrictions on the number of poles and the latter were eliminated by Wüstholz [260] to give the unconditional result cited above.

We now address the question of when a linear form in the numbers appearing in Theorem 6.5 can actually vanish. Let r be the rank of the subgroup of $\mathbb{C}/\mathbb{Z}\omega$ generated over the rationals by $u_1, \ldots, u_n$; then the question is answered by the following result.

Theorem 6.7 *The vector space generated over* $\overline{\mathbb{Q}}$ *by*

$$1,\ \omega,\ \eta,\ \lambda(u_1, \omega), \ldots, \lambda(u_n, \omega)$$

has dimension $3 + r$.

For the proofs of the results we refer to [260]. They can also be obtained from the analytic subgroup theorem on the lines already indicated. We mention that it is possible to supplement the set of numbers considered in Theorem 6.7 by $u_1, \ldots, u_n$ and we find that the dimension of the vector space generated by the expanded set is then $3 + 2r$. Finally we remark that all these results can be extended to arbitrary abelian varieties; the statements become much more involved and we shall not consider them in detail here. However, we shall come back to the general situation in the next section where we shall discuss a conjecture of Leibniz.

6.3 Transcendence properties of rational integrals

Here we study transcendence properties of rational integrals on projective varieties. These investigations were motivated by the fact that many numbers which have in the past been established as transcendental can be written as rational integrals with algebraic bounds. The results can now be incorporated as special cases of a general theorem, namely Theorem 6.8 below. Another source of the investigations is a very interesting monograph of Arnol'd [8] where there is a reference to a letter from Leibniz to Huygens [113] dated 1691 (actually dated $\frac{10}{20}$ April in the style of the time). In his letter, Leibniz formulated the problem of transcendence of the areas enclosed by segments of an algebraic curve and two straight lines, where all the defining equations have rational coefficients. Arnol'd [8, p. 105] rephrased the problem in modern language: he asked whether an abelian integral along an algebraic curve with rational (algebraic) coefficients taken between limits which are rational (algebraic) numbers is generally a transcendental number. Leibniz' letter is of interest historically for it shows that there was some concept of transcendence as early as the seventeenth century, long before the work of Liouville, Hermite and Lindemann when the subject is traditionally said to have been initiated.

The problem of Leibniz, as formulated by Arnol'd, can be extended by asking whether the number

$$I(\xi,\gamma) = \int_\gamma \xi$$

is, in general, transcendental; here ξ is a closed holomorphic 1-form on a smooth quasi-projective variety X defined over a number field, so that $\xi \in \Gamma(X, \Omega^1_X)^{d=0}$, and γ is a (continuous) path mapping $[0, 1]$ to $X(\mathbb{C})$ such that $\gamma(0)$, $\gamma(1)$ are in $X(\overline{\mathbb{Q}})$. Note that $I(\xi,\gamma)$ depends only on the homology class $[\gamma]$ of the path γ and not on the path itself, and, moreover, only on the cohomology class of the differential form ξ up to an additive algebraic number. We observe at once that $I(\xi,\gamma)$ is not always transcendental; for instance, the differential form ξ could be exact and then the integral obviously takes algebraic values. Furthermore we can give non-trivial counterexamples as follows. Let E be an elliptic curve defined over a number field and let $\Gamma \subseteq E \times E$ be the graph of an endomorphism φ of E; then the differential form $\xi = \mathrm{pr}_1^* \varphi^* dx/y - \mathrm{pr}_2^* dx/y$, with the usual notation regarding pull-back etc., vanishes on Γ so that, when the path γ is contained in Γ, the integral $I(\xi,\gamma)$ is zero. Essentially these are the only exceptional cases for we have the following basic theorem.

Theorem 6.8 (Integral theorem) *There exist an algebraic cycle Z and a subgroup Ω of the integral homology of X, both depending only on the differential form ξ, such that if $I(\xi,\gamma)$ is algebraic then we have $\gamma(0), \gamma(1) \in Z(\overline{\mathbb{Q}})$ and $\int_{\gamma+\omega} \xi \in \overline{\mathbb{Q}}$ for all $\omega \in \Omega$.*

Both the cycle Z and the subgroup Ω can be determined effectively; they arise from certain algebraic subgroups of group varieties attached to the pair X, ξ. Thus if, for example, we assume that $[\gamma] \neq 0$ and ξ is not exact and moreover that X is projective and the associated Albanese variety of X is simple then we deduce that $I(\xi,\gamma)$ is transcendental. In fact in this case it turns out from the proof of Theorem 6.8 that the subgroup Ω is trivial. As a further immediate consequence of the proof of the integral theorem we obtain the following corollary.

Corollary 6.9 *Non-zero periods of rational integrals are transcendental.*

Plainly Corollary 6.9 generalises Corollary 6.6 discussed in the previous section. Thus, in particular, it implies that π is transcendental; indeed we have $2\pi i$ as a period of the differential form dx/x. It includes also the classical result of Schneider [211, II] on the transcendence of the circumference of an ellipse with algebraic axes lengths; this was an immediate consequence of his result on ω and η discussed in Section 6.2. Moreover, it gives the generalisations of the latter by Baker [18] and by Wüstholz [261] implying the transcendence of the sum of the circumferences of several such ellipses. These results show that not only is it impossible to square the circle in the classical Greek fashion but also it is, in a sense, impossible to square the ellipse; that is, we cannot construct with ruler and compasses only a square with circumference equal to that of any given ellipse with algebraic axes lengths. In view of Corollary 6.9, this result can now be generalised to an arbitrary real algebraic projective curve and it follows that, if the latter is defined over an algebraic number field and the differential giving the length is not exact, then the perimeter of the curve is transcendental.

It will be observed that the theorem and corollary include several other classical results as special cases. In particular we obtain the transcendence of e^α for non-zero algebraic α due to Lindemann; it suffices to take the integral of the differential form dx/x between the limits 1 and e^α. Further, we have the fundamental results of Schneider [214] on abelian integrals with algebraic bounds and, in particular, his well-known theorem, already referred to in Section 2.3, on the transcendence of the Beta-function

$$B(a,b) = \frac{\Gamma(a)\,\Gamma(b)}{\Gamma(a+b)} = \int_0^1 x^{a-1}(1-x)^{b-1}dx$$

with rational arguments a, b such that a, b and $a+b$ are non-integral; the result is immediate from the fact that the Beta-values are periods of differentials of the second kind on Fermat curves and indeed here the Jacobians are of CM-type. In another direction, the Gelfond–Schneider theorem can be deduced from Theorem 6.8 on taking X as the affine plane with the axes removed and ξ as a differential of the form $(dx/x) - \beta(dy/y)$. Furthermore we note that the theorem gives at once the transcendence of the integral

$$\int_0^1 \frac{dx}{1+x^3} = \frac{1}{3}\left(\log 2 + \frac{\pi}{\sqrt{3}}\right).$$

Siegel mentioned in his book [231] that it was not known at the time whether the integral was irrational or not and the problem was solved by Baker [15]; in fact the transcendence is an obvious consequence of the fundamental theorem on logarithmic forms.

It may be of interest to make some further remarks on the historical development of the Leibniz problem. As Arnol'd points out in his monograph [8], the origins can be traced back to Kepler's discoveries in celestial mechanics. Newton was greatly interested in Kepler's work and, through Kepler's second law, he was led to investigate whether the solution x of the equation $x - e \sin x = t$ is an algebraic or transcendental function of the time t. The outcome was formulated as Lemma XXVIII in his *Principia* and states

> *Nulla extat figura Ovalis cujus area rectis pro lubitu abscissa possit per aequationes numero terminorum ac dimensionum finitas generaliter inveniri.*

Thus it would seem that Newton was claiming to have proved a special case, relating to 'oval figures', of the problem referred to at the beginning. The lemma was the starting point of a long discussion as to whether its proof was correct or not. In fact Leibniz in his letter to Huygens gave as a counterexample the Bernoulli lemniscate with equation $a^2x^2 = y^2(a^2 - y^2)$. The relevant area enclosed by the lemniscate is given by

$$\int_y^a x\, dy = \frac{1}{3a}(a^2 - y^2)^{3/2}$$

and thus is an algebraic function of y; indeed this is clear from the fact that the 1-form which calculates the area is exact. In his further discussion Leibniz then raises a question which amounts to the transcendence of the values of abelian integrals at an algebraic point as formulated in Arnol'd's book. To our knowledge it is one of the first places where there is an allusion to the concept of transcendence and where a general question in the area is formulated explicitly. Of course, the ancient Greek problem of 'squaring the circle' was very fashionable at the time and the first attempts to disprove it were made by J. Gregory in 1667 (see [55, p. 421]). By 1775 the circle-squarers had become so numerous that the Academy of Sciences in Paris passed a resolution that no purported solution would be officially examined. Lambert [133] succeeded in showing in 1761 that π is irrational and, in 1882, Lindemann [146] established the transcendence of π and thus finally solved the quadrature problem.

6.4 Algebraic groups and Lie groups

Preliminary to the proof of the integral theorem, we give an exposition of some relevant facts on commutative algebraic groups and their associated complex Lie groups. Let G be a connected algebraic group defined over an algebraically closed field and let $\mathfrak{g}$ be its associated Lie algebra. We assume that the field of definition is the algebraic closure $\overline{\mathbb{K}}$ of some algebraic number field $\mathbb{K}$; this will simplify our discussion though the assumption is not strictly necessary. Let $G(\mathbb{C})$ be the complex Lie group of G and let $\mathfrak{g}(\mathbb{C})$ be its corresponding complex Lie algebra. Then given any tangent vector $X \in \mathfrak{g}(\mathbb{C})$ there exists a unique analytic homomorphism $\varphi_X : \mathbb{G}_a(\mathbb{C}) \to G(\mathbb{C})$ such that

$$(d\varphi_X)\Big(\frac{d}{dt}\Big) = X\,.$$

This property is used to construct the exponential map of the Lie group $G(\mathbb{C})$ in the following way: for every $X \in \mathfrak{g}(\mathbb{C})$ we define

$$\exp_G(X) = \varphi_X(1)$$

and we get an analytic homomorphism $\exp_G$ from the Lie algebra $\mathfrak{g}(\mathbb{C})$ of $G(\mathbb{C})$ into $G(\mathbb{C})$.

Now we observe that any analytic homomorphism $\varphi\colon \mathbb{G}_a(\mathbb{C}) \to G(\mathbb{C})$ induces a homomorphism $\delta\varphi\colon \mathfrak{g}^*(\mathbb{C}) \to \mathfrak{g}_a^*(\mathbb{C})$ between the spaces dual to the Lie algebras as well as a homomorphism $d\varphi$ between the Lie algebras themselves. If (,) is the pairing $\mathfrak{g} \times \mathfrak{g}^* \to \overline{\mathbb{K}}$ which defines duality then we have

$$\left(\frac{d}{dt}, (\delta\varphi)(\omega)\right) = \left((d\varphi)\left(\frac{d}{dt}\right), \omega\right).$$

Another pairing is obtained by taking $\gamma\colon [0,1] \to G(\mathbb{C})$ to be any path in $G(\mathbb{C})$ and integrating a closed differential form along the path; this defines an element $I(\gamma)$ in $\mathfrak{g}(\mathbb{C})$ by

$$I(\gamma)(\omega) = \int_\gamma \omega = \langle \gamma, \omega \rangle.$$

It is clear from Stokes' theorem that $I(\gamma)$ depends only on the homology class of the path γ. For any element X in $\mathfrak{g}(\mathbb{C})$, let $\gamma_X : [0,1] \to G(\mathbb{C})$ be the path obtained by restricting the analytic homomorphism φ_X to the interval $[0,1]$. Then the following lemma holds.

Lemma 6.10 *We have* $I(\gamma_X) = X$.

Proof. By the transformation formula for integrals we have

$$I(\gamma_X)(\omega) = \int_{\gamma_X} \omega = \int_0^1 (\delta\gamma_X)(\omega) = \left(\frac{d}{dt}, (\delta\gamma_X)(\omega)\right).$$

But the latter is equal to

$$\left((d\gamma_X)\left(\frac{d}{dt}\right), \omega\right) = (X, \omega),$$

and we find therefore that

$$I(\gamma_X)(\omega) = (X, \omega),$$

which means $I(\gamma_X) = X$ as required. □

If X is an element of the kernel of the exponential map $\exp_G$ then $\gamma_X(1) = 0$ whence γ_X is a cycle and so, by the lemma, X is a period vector. Further, if γ is a path in $G(\mathbb{C})$ then there is a unique path of the form γ_X in the homotopy class of γ; thus the integrals $I(\gamma)$ and $I(\gamma_X)$ have the same value and hence

$$\gamma(1) = \gamma_X(1) = \exp_G(X) = \exp_G\bigl(I(\gamma)\bigr).$$

This leads to the following corollary.

Corollary 6.11 *If P is any point in $G(\mathbb{C})$ and γ is any path from 0 to P then*

$$\exp_G\bigl(I(\gamma)\bigr) = P.$$

In view of this corollary, we see that the lemma implies that integration is inverse to exponentiation as one would expect.

Suppose now that G is commutative. Then it is well known that G is an extension of an abelian variety by a linear algebraic group which can be written as a product of a torus $\mathcal{T}$ with a unipotent group V. The torus $\mathcal{T}$ can be written in the form $\mathbb{G}_m^u$ and the group V takes the form $\mathbb{G}_a^v$. For any invariant 1-form ω on G, we denote by H the largest connected algebraic subgroup of G such that ω vanishes on H. Then there exists an invariant 1-form ω' on the quotient group G/H such that ω is the pull-back of ω' and we write $G(\omega) = G/H$. More generally, let X be a smooth quasi-projective variety over $\overline{\mathbb{Q}}$ and let ξ be a closed holomorphic 1-form on X. Then, by Satz 6 in [92], there exists a connected commutative algebraic group G, an open subvariety U of X, a morphism $\varphi\colon U \to G$ and an invariant differential form ω on G such that $\xi = \varphi^*(\omega)$. If $\pi : G \to G(\omega)$ is the canonical projection then the composition $\varphi(\xi)$ of the maps φ and π is a regular map from U into $G(\omega)$. Since $G(\omega)$ depends only on the original differential form ξ we may write $G(\xi)$ in place of $G(\omega)$. We have therefore proved the following lemma.

Lemma 6.12 *Let ξ be a closed holomorphic 1-form on a smooth quasi-projective variety X. Then there exists an open subset U of X, a connected commutative algebraic group $G(\xi)$, a regular map $\varphi(\xi)\colon U \to G(\xi)$ and an invariant differential form ω on $G(\xi)$ such that $\xi = (\varphi(\xi))^*(\omega)$.*

Let G be a commutative algebraic group with linear part L and let i be the canonical injection of L into G. For any homomorphism α from

L to another commutative linear algebraic group L' we define a group $\alpha_* G$ as the push-out of the group G with respect to the morphisms i and α. In other words, $\alpha_* G$ is the quotient of the product $G \times L'$ by the group L which acts on the first factor via i and on the second factor via α, and it is an extension of the abelian part A of G by the linear group L'. The group G injects into the group $G \times L'$ which itself projects onto the group $\alpha_* G$ and we denote the composition of the two maps by β; thus the map β is the homomorphism from G to $\alpha_* G$ induced by α.

It can be seen easily that the unipotent part $V(\xi)$ of the group $G(\xi)$ defined in the lemma has dimension at most 1. Indeed otherwise there exists an algebraic subgroup V of $V(\xi)$ such that the differential form ω vanishes identically on V, whence there exists a non-zero homomorphism α from $V(\xi)$ to $\mathbb{G}_a$ with kernel V. The homomorphism α induces a projection β from $G(\xi)$ to $\alpha_* G(\xi)$ with the same kernel V and, since the dimension of V is positive, the dimension of $\alpha_* G(\xi)$ is strictly less than that of $G(\xi)$. Further, the differential form ω descends to a differential form η on $\alpha_* G(\xi)$ such that $\omega = \beta^*(\eta)$ and, as a consequence, we have $\xi = (\beta \circ \varphi(\xi))^*(\eta)$ which contradicts the lemma.

6.5 Lindemann's theorem for abelian varieties

In Section 1.2 we discussed Lindemann's theorem on the algebraic independence of $e^{\alpha_1}, \ldots, e^{\alpha_n}$ where $\alpha_1, \ldots, \alpha_n$ are algebraic numbers linearly independent over $\mathbb{Q}$. As described in Section 1.3 the work motivated Siegel to introduce the class of E-functions extending the classical exponential function and he then proceeded to establish an important generalisation of Lindemann's theorem in this context. We recall that the coefficients in the Taylor expansion of an E-function have the form $a_n/n!$ where the algebraic numbers a_n and their denominators are bounded exponentially in terms of n and so are dominated by the $n!$. Further we recall that, in order to obtain his basic result on the algebraic independence of E-functions, Siegel made the hypothesis that the functions in question satisfy a system of linear differential equations over the field of rational functions of a single variable. This implies that the vector space generated by the solutions over the rational function field is invariant under differentiation and thus it allows one to utilise Wronskians for linear elimination. Again as described in Section 1.3 Siegel was able to deal only with linear differential equations of second order and it was Shidlovsky who succeeded in establishing the desired result for equations of arbitrary order.

The Baker–Wüstholz theory also has its origins in the classical exponential function but it puts it now into the geometric framework of commutative algebraic groups. We consider the algebraic torus $\mathcal{T}$ of dimension n which is basic for the Baker theory and we take $X_1(\partial/\partial X_1), \ldots, X_n(\partial/\partial X_n)$ as a basis for its Lie algebra (see Section 4.4). The tangent vector

$$\partial_\alpha = \alpha_1 X_1(\partial/\partial X_1) + \cdots + \alpha_n X_n(\partial/\partial X_n) \in \mathrm{Lie}(\mathcal{T})$$

has image $\exp_{\mathcal{T}}(\partial_\alpha) = (e^{\alpha_1}, \ldots, e^{\alpha_n})$ in $\mathcal{T}$ (see Lemma 6.10 together with Corollary 6.11) and Lindemann's theorem is then equivalent to the assertion that the dimension of the Zariski closure $\overline{g_\alpha}$ of $g_\alpha = \exp_{\mathcal{T}}(\partial_\alpha)$ in $\mathcal{T}$ taken over the field $\overline{\mathbb{Q}}$ is equal to $\dim \mathcal{T}$. In other words the point g_α is dense in $\mathcal{T}$ with respect to the Zariski topology.

The main problem in this field is to obtain a natural extension of the classical Lindemann theorem in the context of abelian varieties and more generally to put the Siegel–Shidlovsky theorem into a geometrical framework. The most significant result to date relates to elliptic curves; it was published in 1983 in separate papers in the same issue of *Inventiones Mathematicae* by Philippon [192] and by Wüstholz [259] and asserts as follows.

Theorem 6.13 *Let E be an elliptic curve with complex multiplication by an imaginary quadratic field k and let $\alpha_1, \ldots, \alpha_n$ be algebraic numbers which are linearly independent over k. Then the numbers*

$$\wp(\alpha_1), \ldots, \wp(\alpha_n)$$

are algebraically independent over the field $\overline{\mathbb{Q}}$.

The case $n=1$ of Theorem 6.13 was established by Schneider [213] and indeed without any assumption on the endomorphism algebra of the elliptic curve; in fact, as remarked in Section 2.3, Schneider showed in 1937 that if $\wp(z)$ is a Weierstrass $\wp$-function with algebraic invariants then $\wp(\alpha)$ is transcendental for algebraic $\alpha \neq 0$. However, the first breakthrough towards the general assertion was achieved by Chudnovsky [68] in 1980; he verified Theorem 6.13 in the cases $n \leq 3$.

The proofs of Theorem 6.13 are discussed in the original papers in the wider context of a general abelian analogue of Lindemann's theorem and we shall follow that approach here. Accordingly let A be a simple abelian variety defined over $\overline{\mathbb{Q}}$ and let $k = (\operatorname{End} A) \otimes_{\mathbb{Z}} \mathbb{Q}$. The abelian analogue of the torus $\mathcal{T}$ referred to above is the algebraic group $\mathbb{A} = A^n$ which has a ring of endomorphisms $\operatorname{End} \mathbb{A}$ isomorphic to the ring of $n \times n$ matrices $\mathcal{M}_n(\operatorname{End} A)$ with elements in $\operatorname{End} A$. The differential of an endomorphism of any algebraic group acts linearly on the Lie algebra of the group and, in particular, this gives an action of the endomorphism algebra $(\operatorname{End} \mathbb{A}) \otimes k$ on the Lie algebra $\operatorname{Lie} \mathbb{A}$ of $\mathbb{A}$ where, as later, the tensor product is with respect to $\operatorname{End} A$. We consider a subfield $K \subseteq (\operatorname{End} \mathbb{A}) \otimes k$ and choose an eigenvector ∂ for the action of K on $\operatorname{Lie} \mathbb{A}$. Then it suffices for Theorem 6.13 to establish the lower bound

$$2 \dim \overline{g} \geq [K : \mathbb{Q}]$$

for the dimension of the Zariski closure $\overline{g}$ of $g = \exp_{\mathbb{A}}(\partial)$. Indeed it is well known that the degree of the skew field k divides $2 \dim A$ with equality if and only if A has complex multiplication by an order in a CM-field. Thus, if $[K : k] = n$, then, for a CM-field k, we get $\overline{g} = \mathbb{A}$ and this can be interpreted as the abelian analogue of Lindemann's theorem.

We verify the latter assertion for an elliptic curve $Y^2 = 4X^3 - g_2X - g_3$, say E, with complex multiplication. There exists a natural basis ∂_E for the Lie algebra of E such that the exponential map $\exp_E$ of E satisfies

$$(\exp_E^* X)(z\partial_E) = X(\exp_E(z\partial_E)) = \wp(z)$$

with respect to this basis. Let $\alpha_1, \ldots, \alpha_n$ be algebraic numbers linearly independent over k and let K be the smallest field containing k and all the α_i. We extend the k-linearly independent set $\alpha_1, \ldots, \alpha_n$ to a basis $\alpha_1, \ldots, \alpha_m$ for K over k so that $m = [K : k]$. The extension K is a vector space over k on which an element $\alpha \in K$ acts by left multiplication as a k-linear endomorphism. The endomorphism α can be expressed in terms of the basis by a matrix $M(\alpha)$ in the matrix algebra $\mathcal{M}_m(k)$ and this shows that K is isomorphic to a subfield of $(\operatorname{End} \mathbb{E}) \otimes k$ of dimension m over k for $\mathbb{E} = E^m$. Since K acts on $\operatorname{Lie} \mathbb{E}$ by left multiplication it follows that the vector $\partial = \alpha_1\partial_1 + \cdots + \alpha_m\partial_m$ is an eigenvector for the action of K. The basic equation $\overline{g} = \mathbb{A}$ discussed above with $g = \exp_{\mathbb{E}}(\partial)$ becomes $\overline{g} = \mathbb{E}$. This implies that $P(g) \neq 0$ for all $P \neq 0$ in the field of rational functions $\overline{\mathbb{Q}}(\mathbb{E})$ on $\mathbb{E}$. We have seen that the coordinate function X_i on the ith factor of $\mathbb{E}$ and the associated coordinate functions z_i on the Lie algebra are related by

$$(\exp_{\mathbb{E}}^* X_i)(z_i\partial_i) = X_i(\exp_{\mathbb{E}}(z_i\partial_i)) = \wp(z_i).$$

Now if $P = P(X_1, \ldots, X_m)$ is a non-zero polynomial in the variables $X_1, \ldots, X_m$ then P is in $\overline{\mathbb{Q}}(\mathbb{E})$ whence we obtain

$$\begin{aligned} P(g) &= P(X_1(g), \ldots, X_m(g)) \\ &= P((\exp_{\mathbb{E}}^* X_1)(\partial), \ldots, (\exp_{\mathbb{E}}^* X_m)(\partial)) \\ &= P(\wp(\alpha_1), \ldots, \wp(\alpha_m)). \end{aligned}$$

Since $P(g) \neq 0$ this implies that $\wp(\alpha_1), \ldots, \wp(\alpha_m)$, whence a fortiori $\wp(\alpha_1), \ldots, \wp(\alpha_n)$, are algebraically independent and this gives Theorem 6.13.

The key ingredients in the proofs of the fundamental lower bound $[K : \mathbb{Q}]$ for $2 \dim \overline{g}$ in the original papers are the classical construction of an auxiliary polynomial by way of Siegel's lemma as in earlier chapters, an interpolation procedure using a Schwarz lemma and an appeal to the theory of multiplicity estimates relating to group varieties. Another essential ingredient is an effective Hilbert Nullstellensatz which, as shown in [259], can be obtained relatively simply from the theory of resultants dating back to Lasker, Noether and Macaulay. The underlying group in both instances is $G = \mathbb{G}_a \times \mathbb{A}$ but since one has to deal with transcendental extensions of algebraic number fields rather than the fields themselves one needs for the application of the Siegel lemma an idea which was introduced into transcendence theory by Philippon [192] and involves in this context taking a large power $\mathbb{G} = G^h$ as the basic group. As a consequence the interpolation procedure has to be performed with respect to functions of h variables and in the multiplicity estimates there are h differential operators as described in Section 5.2. One further aspect critical to the discussion is a transcendence device dating back to Baker and Coates which was subsequently supplemented and utilised in the elliptic case by Anderson [4] and by Chudnovsky [68]; it provides estimates, in the general framework of algebraic groups, for the height and for the degree of the auxiliary polynomial that are crucial to the demonstration. For details of all this work see [192] and [259].

Finally we remark that Brownawell and Kollar (see [131]) have obtained a result on the Hilbert Nullstellensatz for affine varieties analogous to the version for projective varieties referred to above. This would appear to be of value in connection with open problems in this area, in particular the question of a new demonstration of the classical Lindemann theorem adapted from the proofs of Theorem 6.13.

6.6 Proof of the integral theorem

The integral theorem is obtained as an application of the analytic subgroup theorem which we recall states that, in general, an analytic subgroup of a commutative algebraic group has only trivial algebraic points. We begin by constructing the algebraic group G and the analytic subgroup B mentioned in Theorem 6.1. Accordingly let $G(\xi)$ be the algebraic group described in Lemma 6.12. The standard differential forms ω on $G(\xi)$ and dx on $\mathbb{G}_a$ pull back to invariant differential forms η and dt on the group $G = G(\xi) \times \mathbb{G}_a$ and we define a new invariant differential form τ on this group as $\eta - I(\xi, \gamma)dt$. Now assuming that $I(\xi, \gamma)$ is algebraic, it follows that τ is an invariant differential form on G. Furthermore, on signifying by $\mathfrak{g}$ the Lie algebra of G and by $\mathfrak{g}^*$ its dual and noting that τ is in $\mathfrak{g}^*$ we can take B as the analytic subgroup of the Lie subalgebra of $\mathfrak{g}$ defined by the condition $\tau = 0$.

By Lemma 6.10, the functional $I(\gamma)$ on the space of invariant differential forms is an element of the Lie algebra of $G(\xi)$ and therefore the mapping $d\varphi(\xi)$ tangent to $\varphi(\xi)$ takes $I(\gamma)$ to an element u of the Lie algebra of the complex Lie group $G(\xi)(\mathbb{C})$. Let i_1 and i_2 be the canonical injections of $G(\xi)$ and $\mathbb{G}_a$ into $G(\xi) \times \mathbb{G}_a$ and define the element ε of the Lie algebra $\mathfrak{g}(\mathbb{C})$ by

$$\varepsilon = di_1(u) + di_2(d/dx).$$

Then it is easily seen that ε is contained in the Lie algebra $\mathfrak{b}(\mathbb{C})$ of the analytic subgroup B. As we have shown in Corollary 6.11, the image

P_1 of u in $G(\xi)(\mathbb{C})$ under the exponential map $\exp_G$ is actually an element of $G(\xi)(\overline{\mathbb{Q}})$. Further, the image P_2 of the element d/dx of the Lie algebra of $\mathbb{G}_a$ under the exponential map $\exp_{\mathbb{G}_a}$ is in $\overline{\mathbb{Q}}$. It follows that $P = i_1(P_1)+i_2(P_2)$ is in $G(\overline{\mathbb{Q}})$. On the other hand we have $P = \exp_G(\varepsilon)$ whence P is also in $B(\overline{\mathbb{Q}})$.

Now from Theorem 6.1 there exists an algebraic subgroup Γ of G such that $\varepsilon \in \operatorname{Lie}\Gamma \subseteq \mathfrak{b}$. Further, by definition, B is the graph of an analytic homomorphism $\psi: G(\xi) \to \mathbb{G}_a$ whence the algebraic subgroup Γ induces by restriction a homomorphism $\psi_H: H \to \mathbb{G}_a$ where H is the projection of Γ onto the first factor $G(\xi)$ of G. Since τ vanishes on $\mathfrak{b}$ we see that it vanishes on $\operatorname{Lie}\Gamma$ and this implies that ω vanishes on the kernel of ψ_H. By the discussion preceding Lemma 6.12, $G(\xi)$ does not have a non-trivial algebraic subgroup on which ω vanishes. This shows that ψ_H is injective and that H is contained in the largest unipotent subgroup of $G(\xi)$ which is either trivial or equal to $\mathbb{G}_a$. We now define $Z = \varphi(\xi)^{-1}H$ and $\Omega = \varphi_*^{-1}[0]$ where $\varphi_*: H_1(X,\mathbb{Z}) \to H_1(G(\xi),\mathbb{Z})$ is the group homomorphism induced by φ on the homology groups and this completes the proof of Theorem 6.8.

6.7 Extended multiplicity estimates

In this section we generalise the theory developed in Chapter 4 to arbitrary commutative group varieties. We fix, as usual, an algebraically closed field $\mathbb{K}$ of characteristic zero and consider a group variety G over $\mathbb{K}$ with dimension n which we assume to be embedded in $\mathbb{P}^N$. We begin by defining two types of operators similar to those in Chapter 4, namely translation and differential operators, the latter being obtained from the former.

We recall that in Section 6.1 we defined addition formulae on an arbitrary group variety according to which there exist positive integers a, b and a finite covering of $G \times G$ by open sets $\mathcal{U}$ with the following property. For each $\mathcal{U}$ there exists a set of bihomogeneous polynomials $P_0(X',X''),\ldots,P_N(X',X'')$ in the variables $X' = (X'_0,\ldots,X'_N)$, $X'' = (X''_0,\ldots,X''_N)$ of bidegree a, b such that

$$(X_0(g+h) : \ldots : X_N(g+h)) = (P_0(g,h) : \ldots : P_N(g,h))$$

for all $g, h \in \mathcal{U}$. We refer to Section 6.9 where we explain how such a complete set of addition formulae can be effectively constructed.

We shall now define operators analogous to those introduced in Chapter 5. In the present situation the basic algebra R associated with G is the quotient of the polynomial ring in the variables $X_0, \ldots, X_N$ by the homogeneous ideal generated by the collection of polynomials vanishing on G. The natural graduation of the polynomial ring by degree induces a graduation $R = \bigoplus_{k \geq 0} R_k$ on R with the property that R is generated as a ring by the subspace R_1 of elements of degree 1. For g in $G(\mathbb{K})$, that is the set of points rational over $\mathbb{K}$, we let S_g be the local ring of G at g consisting of all rational functions, in other words all elements in the quotient field of R which are regular at g. The graduation on R induces naturally a graduation $\bigoplus_{k \geq 0} S_g \otimes R_k$ on $S_g \otimes R$.

Addition on G induces a finite collection of homomorphisms $\mu \colon R \to S_g \otimes R$ such that if $r_0, \ldots, r_N$ is a basis for R_1 over $\mathbb{K}$ then the projective coordinates of $g + h$ satisfy

$$(r_0(g+h) : \ldots : r_N(g+h)) = (\mu(r_0)(g,h) : \ldots : \mu(r_N)(g,h)) \quad (6.1)$$

and this holds provided that the $\mu(r_i)$, $0 \leq i \leq N$, are not all zero at (g, h). By our choice of the projective space there exists at least one μ for which this is satisfied. It is easy to see that if μ_1 and μ_2 are addition laws then $\alpha_1\mu_1 + \alpha_2\mu_2$ is an addition formula for all α_1, α_2 in $\mathbb{K}$. This implies that for any countable collection of points we may choose α_1 and α_2 such that formula (6.1) is valid.

For each $g \in G$ the symmetric algebra $\mathrm{Sym}(\mathfrak{g})$ of the Lie algebra of G acts on the local ring S_g as an algebra of differential operators and each ∂ in $\mathrm{Sym}(\mathfrak{g})$ defines an endomorphism $\mathcal{D}$ of $S_g \otimes R$ given by $s \otimes r \mapsto \partial s \otimes r$. The evaluation homomorphism taking $s \mapsto s(g)$ at g, where $s \in S_g$, induces a homomorphism $\gamma \colon S_g \otimes R \to R$ mapping $s \otimes r$ to $s(g)r$. Thus we obtain an endomorphism $r \mapsto \gamma(\mathcal{D}(\mu(r)))$ of R which we shall denote by $T_g \circ \mathcal{D}$; this operator does not depend on the choice of μ. The symmetric algebra $\mathrm{Sym}(\mathfrak{g})$ has a natural graduation so that we may define the order of an element $r \in R$ at $g \in G$ to be the largest integer m such that the element $(T_g \circ \mathcal{D})(r)$ evaluated at $0 \in G$ is 0 for all $\mathcal{D}$ with degree $< m$.

We shall now state, following [263], an analogue for commutative group varieties of the fundamental theorem of Chapter 5. Accordingly, we associate with a commutative group variety G and a subspace V of its Lie algebra $\mathfrak{g}$ an index $\tau(G, V)$ which we define as d/n, where d is the dimension of V. We call V semistable if for all proper quotients $\pi : G \to G'$ we have

$$\tau(G', \pi_*(V)) \geq \tau(G, V)$$

where π_* is the homomorphism induced on $\mathfrak{g}$ by π. It follows that if V is not semistable then there exists a non-trivial quotient $\pi : G \to G^*$ such that $\tau(G^*, \pi_* V)$ is minimal and $\tau(G^*, \pi_* V) < \tau(G, V)$. Clearly $\pi_* V$ is semistable and it can be shown (see [93]) that G^* is uniquely determined provided that this quotient is selected with minimal dimension.

We fix an element $g \in G$ and a homomorphism μ for which the formula (6.1) is valid for all elements sg in the group generated by g. By the above remarks this is possible and we obtain translation operators T_{sg} as before. We signify by $L_1, \ldots, L_d$ differential operators corresponding to a basis for V, by $r \neq 0$ an element in R_D and we take S, T as positive integers. Then we have the following theorem.

Theorem 6.14 *Suppose that V is semistable. Then there exists an effectively computable constant $c > 0$, depending only on G and V, with the following property. If $ST^d > cD^n$ and if, for all integers s with $0 \leq s \leq S$ and all non-negative integers $t_1, \ldots, t_d$ with $0 \leq t_1 + \cdots + t_d \leq T$, we have*

$$\left(T_{sg} L_1^{t_1} \cdots L_d^{t_d} r\right)(0) = 0$$

then there exists an integer s' with $0 < s' < S$ such that $s'g = 0$.

There are basically two differences between the theorem here and the fundamental Theorem 5.7 on multiplicity estimates. First, in Chapter 5, we dealt with a product of group varieties $G_0 \times \cdots \times G_n$ and an embedding into multiprojective space with $G_0 = \mathbb{G}_a$ and $G_1 = \cdots = G_n = \mathbb{G}_m$. This made it possible to take a multidegree polynomial with respect to the several factors. Here we have only one factor and therefore only one degree. However, it should be remarked that the theorem can easily be extended to cover also the product case. The second difference is that we have assumed that V is semistable. This makes it possible to rule out a priori a conclusion analogous to (i) in Theorem 5.7 and this is why we have here the simple condition $ST^d > cD^n$. It can be easily seen that if (i) is ruled out and if V is semistable with respect to the subspace of $\mathfrak{g}$ generated by $\Delta_0, \ldots, \Delta_{n-1}$, then (ii) is equivalent to a condition on S, T and D as above. Theorem 6.14 above has a simpler form than the fundamental theorem on multiplicity estimates but it is not so universal; nevertheless it is sufficient to prove the analytic subgroup theorem.

The proof of Theorem 6.14 runs along the same lines as that of the earlier theorem. We put

$$S^{(r)} = ((n - r + 1)/n)S, \quad T^{(r)} = ((n - r + 1)/n)T \quad (1 \le r \le n).$$

Then, assuming that the theorem is false, it is proved by induction on r that the ideal I_r generated by the homogeneous elements

$$T_{\sigma\gamma} L_1^{\tau_1} \cdots L_d^{\tau_d} P \quad (0 \le \sigma \le S - S^{(r)},\ 0 \le \tau_1 + \cdots + \tau_d \le T - T^{(r)})$$

has rank r and that the set of polynomials

$$T_{s\gamma} L_1^{t_1} \cdots L_d^{t_d} I_r \quad (0 \le s \le S^{(r)},\ 0 \le t_1 + \cdots + t_d \le T^{(r)})$$

vanishes at 0; here essential use is made of the hypothesis that V is semistable. Now the ideal I^* generated by the ideals $T_{s\gamma} I_n$ with $0 \le s \le S^{(n)}$ has rank n and the elements in each of the ideals in the set

$$L_1^{t_1} \cdots L_d^{t_d} I^* \quad (0 \le t_1 + \cdots + t_d \le T^{(n)})$$

vanish at $s\gamma$ for $0 \le s \le S^{(n)}$. We now apply degree theory as developed in Chapter 5, in particular the results on the length of ideals referred to there, to conclude that

$$S^{(n)} (T^{(n)})^d \le c' D^n$$

for some effectively computable constant c' provided that the elements $s\gamma$ for $0 < s < S$ are pairwise distinct; on taking $c = n^2 c'$ we obtain a contradiction and the theorem follows at once.

It will be observed that Theorem 6.14 has a slightly simpler shape than the main theorem in [263]. To get the result stated here from the latter, one applies semistability to verify the principal conditions as recorded on [263, p. 475]; the deduction is in fact an easy exercise.

6.8 Proof of the analytic subgroup theorem

The proof of Theorem 6.1 divides naturally into two parts. The first is the so-called constructive part and it follows the usual pattern in connection with logarithmic forms as described in Chapter 2. However, the underlying group G here is more general than the linear algebraic group $\mathbb{G}_a \times \mathbb{G}_m^n$ occurring in the original work and this introduces several technical difficulties. The second is the deconstructive part and this is treated by Theorem 6.14 on multiplicity estimates on group varieties. The argument we give below is a modified version of that of Wüstholz [264] and we refer there for further details.

Let G be a commutative group variety with Lie algebra $\mathfrak{g}$ and subalgebra $\mathfrak{b}$ defined over the algebraic closure $\overline{\mathbb{Q}}$ of $\mathbb{Q}$. We obtain from $\mathfrak{b}$ via the exponential map an analytic subgroup B of $G(\mathbb{C})$; we say that B is semistable if $\mathfrak{b}$ is semistable in the sense of Section 6.7. In fact, to apply the multiplicity estimates, we shall need B to be semistable, and we proceed now to show that we may reduce Theorem 6.1 to this special case.

Accordingly, suppose that B is not semistable. Then, by Section 6.7, therc exists a proper quotient $\pi : G \to G^*$ of G, defined over $\overline{\mathbb{Q}}$, such that $\tau(G^*)$ is minimal and $\tau(G^*) < \tau(G)$. Since $\tau(G) \leq 1$, we see that $\dim(\pi_*\mathfrak{b}) < \dim(\mathfrak{g}^*)$ whence the image B^* of B in $G^*(\mathbb{C})$ is a proper analytic subgroup of G^*. We now argue by induction on the dimension of the group G; plainly Theorem 6.1 is valid when the dimension is 1 and we assume the validity for groups with dimension smaller than that of G. Now suppose that $B^*(\overline{\mathbb{Q}}) \neq 0$. We can therefore apply the inductive hypothesis to B^* and G^* and we obtain an algebraic subgroup K^* of G^* contained in B^*. The inverse image K of K^* in G properly contains H, whence $\dim H < \dim K$, and thus we have

$$\begin{aligned}\tau(G^*) &= (\dim B - \dim H)/(\dim G - \dim H)\\ &> (\dim B - \dim K)/(\dim G - \dim K) = \tau(G/K) = \tau(G^*/K^*).\end{aligned}$$

This contradicts the minimality of $\tau(G^*)$. We conclude that $B^*(\overline{\mathbb{Q}}) = 0$ and hence $B(\overline{\mathbb{Q}}) = (B \cap \ker \pi)(\overline{\mathbb{Q}})$. A second application of the inductive hypothesis applied now to $B \cap \ker \pi$ and $\ker \pi$ gives an algebraic subgroup of G contained in B and Theorem 6.1 follows.

Henceforth, therefore, we can assume that B is semistable. Since furthermore G and $\mathfrak{b}$ are defined over the algebraic closure of $\mathbb{Q}$, there exists a number field $\mathbb{K}$ such that both G and $\mathfrak{b}$ are defined over $\mathbb{K}$. Thus it will suffice to prove the following result.

Theorem 6.15 (Semistability theorem) *Let G be a commutative group variety and let B be a proper analytic subgroup of $G(\mathbb{C})$ with both B and G defined over a number field $\mathbb{K}$. If B is semistable then $B(\overline{\mathbb{K}}) = 0$.*

Proof. We fix a very ample line bundle $\overline{L}$ over $\overline{G}$ as in Section 6.7 and obtain an embedding $\varphi\colon \overline{G} \to \mathbb{P}^N$. Further we take $\exp_G$ to be the exponential map from $\mathfrak{g} \otimes \mathbb{C}$ into $G(\mathbb{C})$ and we write $f_i = X_i(\varphi(\exp_G))$ for $i = 0, \ldots, N$, where $X_0, \ldots, X_N$ are the homogeneous coordinates for $\mathbb{P}^N$. The functions are defined on $\mathfrak{g} \otimes \mathbb{C}$ which we identify with $\mathbb{C}^n$ where $n = \dim G$. It is known that

$$\log |f_i(z)| \leq c_1 + c_2 \|z\|^2, \quad i = 0, \ldots, N,$$

for any $z = (z_1, \ldots, z_n)$ in $\mathbb{C}^n$, where $\|z\|^2 = z_1\bar{z}_1 + \cdots + z_n\bar{z}_n$ is the standard metric on $\mathbb{C}^n$ and where c_1, c_2 are positive constants depending only on G and L.

A basic tool in the theory of logarithmic forms is the maximum-modulus principle which has come to be applied in the shape of some version of the so-called Schwarz lemma. Lef f be a function of the complex variable w holomorphic in a closed disc centred at the origin and with radius $r > 0$. We put

$$\|f\|_r = \max_{|w|=r} |f(w)|.$$

Now suppose that f has a zero at each of the points $s = 0, 1, \ldots, S$ with order T and that $0 < S < r' < r$. We consider the finite Blaschke product

$$g(w) = \prod_{s=0}^{S} \left(\frac{r^2 - sw}{r(w - s)} \right)^T.$$

Plainly, the function fg is holomorphic in the disc $|w| \leq r$ and the maximum-modulus principle gives

$$\|fg\|_{r'} \leq \|fg\|_r.$$

Each factor in the product representing $g(w)$ has modulus 1 when $|w| = r$ and modulus at least $(r^2+r'^2)/(2rr')$ when $|w| = r'$ (indeed the typical factor has modulus at least $(r^2 + r's)/(rr' + rs)$ on $|w| = r'$). Hence we obtain

$$\log \|f\|_{r'} \leq \log \|f\|_r + ST \log\left(\frac{2r\,r'}{r^2 + r'^2}\right)$$

and this is the form of the Schwarz lemma that we shall use later.

Before beginning the main constructive part of the proof we remark further that we shall use the absolute logarithmic Weil height $h = h_{\overline{L}}$ associated with the invertible sheaf $\overline{L}$. We recall two basic inequalities relating to the height, namely

$$h(\ell g) \leq c_3 \ell^2 h(g) + c_4\,, \quad h(g) \leq c_5 \ell^{\varkappa}(h(\ell g) + 1)$$

valid for any positive integer ℓ and any $g \in G(\overline{\mathbb{Q}})$, where c_3, c_4, c_5 and $\varkappa$ denote positive numbers depending only on G and $\overline{L}$. Finally, in connection with the Siegel lemma, we need to know the dimension of the space of global sections $H^0(\overline{G}, \overline{L}^D)$, where D is a positive integer. We shall show in Section 6.9, by way of the Riemann–Roch theorem, that $\dim H^0(\overline{G}, \overline{L}) = \chi(\overline{L})$ where $\chi(\overline{L})$ is the Euler characteristic of $\overline{L}$; further, from Section 6.9, $\chi(\overline{L})$ is positive since $\overline{L}$ is ample. Thus, on noting that $\chi(\overline{L}^D) = D^n \chi(\overline{L})$, we have

$$\dim H^0(\overline{G}, \overline{L}^D) = D^n \chi(\overline{L})\,.$$

Now assume that $B(\overline{\mathbb{K}}) \neq 0$ so that B contains a non-trivial algebraic point γ'; we can suppose γ' to be defined over $\mathbb{K}$ by replacing $\mathbb{K}$, if necessary, by a finite extension. Further, we can write $\gamma' = \exp_G(u)$ for some non-zero $u \in \operatorname{Lie} B$. It may happen that γ' is a torsion point and that $\operatorname{ord}(\gamma')$, the order of the group that it generates, is too small to make our argument work. In this case we assume, as clearly we may, that γ' has been selected at the outset so as to be primitive, that is, such that $u \operatorname{ord}(\gamma')$ is primitive in the kernel of the exponential map. We now replace γ' by an element $\gamma = \exp_G(v)$ in B where $v = u/\ell$ and ℓ is a sufficiently large integer. The coordinates of γ are contained in an extension $\mathbb{K}_\ell$ of $\mathbb{K}$ with $[\mathbb{K}_\ell : \mathbb{K}] \leq \ell^{2n}$ and the group Γ generated by γ is contained in $G(\mathbb{K}_\ell)$. Further, the primitive property of γ' implies that $\operatorname{ord}(\gamma) = \ell \operatorname{ord}(\gamma')$.

The first step in the proof of Theorem 6.1 is the construction of a section P of $H^0(\overline{G},\overline{L}^D)$ satisfying

$$(T_{s\gamma'}L_1^{t_1}\cdots L_d^{t_d}P)(0)=0 \tag{6.2}$$

for all integers s with $0\le s\le S$ and all non-negative integers $t_1,\ldots,t_d$ with $0\le t_1+\cdots+t_d\le T$. Here the notation is the same as in Section 6.7 so that d is the dimension of the subspace V defined above, $L_1,\ldots,L_d$ are differential operators corresponding to a basis of V and $T_{s\gamma}$ is a translation operator depending on an open set $\mathcal{U}$; the latter is chosen here so that it contains the points $(0,s\gamma)$ with $0\le s\le \ell S$ whence a fortiori the points $(0,s\gamma')$ with $0\le s\le S$. The quantities S,T,D and ℓ denote large parameters, with S, D, ℓ integral, determined in the course of the argument so as to satisfy a set of inequalities readily seen to be consistent; constants implied by $\ll$ or by $\gg$ will be independent of the parameters. The construction of P is possible since the equations amount to $S'T^d$ linear conditions at most, where S' is the number of distinct points $s\gamma'$ with $0\le s\le S$. Further, on writing P as a homogeneous polynomial in $x_0,\ldots,x_N$, where the latter is a basis for $H^0(\overline{G},\overline{L})$ (so that x_i is the composition of φ and X_i), the number of unknowns in (6.2) is $\dim H^0(\overline{G},\overline{L}^D)$ and, by a critical application of the Baker–Coates device referred to in Section 6.5 so as to eliminate a term TS^2, one sees that the heights of the coefficients are

$$\ll (D+T)\log(D+T)+DS^2 .$$

Hence, by Siegel's lemma together with the result for $\dim H^0(\overline{G},\overline{L}^D)$ recorded above, the system of equations is soluble non-trivially if

$$D^n\chi(\overline{L})\gg S'T^d$$

and, if also $DS^2\ll T$, then, from the first height inequality, there is a solution such that the height of the polynomial representing P is $\ll T\log T$.

We now define a function $\Phi\colon\mathbb{C}\to G(\mathbb{C})$ by $\Phi(w)=\exp_G(wv)$ so that Φ is the one-parameter subgroup determined by v. Note that, since v is not necessarily in $\overline{\mathbb{Q}}^n$, it follows that the tangent space of Φ at the origin may not be defined over $\overline{\mathbb{Q}}$ and here we meet the crux of the Baker method. In order to carry out the customary extrapolation, we take L to be a monomial in the differential operators $L_1, \ldots, L_d$ with degree at most $T/2$ and we put

$$\Psi(w) = \Phi^*(LP),$$

where Φ^* denotes the pull-back induced by Φ. Then, by construction, the function $\psi(w) = \Psi(\ell w)$ has a zero at each $s = 0, \ldots, S$ with order at least $[T/2] + 1$. Further, it is a polynomial in the f_i with degree $\ll D$. By the estimate for these functions cited earlier we obtain

$$\log \|\psi\|_r \ll (D+T)\log(D+T) + D(S^2 + r^2).$$

Hence by the version of the Schwarz lemma derived above with $r' = S$ and $r = S^2$ we conclude that

$$\log \|\psi\|_S \leq -cTS \log S$$

for some $c > 0$ provided that $\log T \ll S$ and $DS^3 \ll T$.

We now consider the number $\Psi(s)$, where s is any integer with $0 \leq s \leq \ell S$, and we proceed to show that $\Psi(s) = 0$. Plainly, from the estimate for ψ above, we have

$$\log |\Psi(s)| \leq -cTS \log S.$$

On the other hand, we observe that $f_i(sv) = \varrho X_i(s\gamma)$ for $i = 0, \ldots, N$ with some non-zero $\varrho = \varrho(sv)$, and thus we have $\Psi(s) = \varrho^\delta \xi$ where $\xi = LP(s\gamma)$ and δ is the degree of LP. Here ξ is an element of $\mathbb{K}_\ell$ and, since $[\mathbb{K}_\ell : \mathbb{K}] \leq \ell^{2n}$, a Liouville-type estimate together with the height inequalities above, shows that, if $\xi \neq 0$, then

$$\log |\xi| \gg -\ell^{2n} h(\xi) \gg -\ell^{2n}(T \log T + \delta(S^2 + \ell^{2+\varkappa})).$$

Another application of the height inequalities gives

$$\log(\max |X_i(s\gamma)|) \ll \ell^{\varkappa} s^2.$$

Further, by [92] and using basic facts about Hermitian forms associated with invertible sheaves (see e.g. Mumford [184]), we obtain

$$\log(\max |f_i(sv)|) \gg -(s^2+1)\|v\|^2.$$

Hence clearly $\log|\varrho| \gg -\ell^{\varkappa}s^2$ and, since $\delta \ll D$, this gives

$$\log|\Psi(s)| \gg -\ell^{2n}(T\log T + \ell^{2+\varkappa}DS^2).$$

We take $D = S^{4d}(\ell'S')$ and $T = S^{4n}(\ell'S')^{(n-1)/d}$ where $\ell' \le \ell$ is a sufficiently large integer so that the condition involving S' in the application of Siegel's lemma is satisfied. Noting that $d \le n-1$ and assuming that $\log S \gg \log\ell$, a comparison of the estimates for $\log|\Psi(s)|$ shows that we have a contradiction. We conclude that $\Psi(s) = 0$ whence it follows from the translation properties of the $T_{s\gamma}$ that (6.2) holds, with γ in place of γ', for all integers $s = 0, \ldots, \ell S$ and all non-negative integers $t_1, \ldots, t_d$ with $t_1 + \cdots + t_d \le T/2$. We now appeal to Theorem 6.14 with S and T replaced by $\ell S'$ and $T/2$. The preliminary supposition on γ is satisfied and, in view of the definition of S' and the primitive property of γ', the order of Γ, if finite, is $\ell S'$ and there is no s' with $s'\gamma = 0$ and $0 < s' < \ell S'$. Finally we have $\ell S'T^d \gg D^n$ and, on assuming that B is semistable, this is clearly inconsistent with Theorem 6.14. Thus we must have $B(\overline{\mathbb{K}}) = 0$ as required. □

6.9 Effective constructions on group varieties

In this section we study some important questions concerning commutative algebraic group varieties. It can be read independently from the material presented so far in this book. However, the reader will be assumed to be familiar with the basic notions in algebraic geometry (see e.g. Hartshorne [125]).

It is known from a classical result of Rosenlicht that a commutative group variety G is an extension of an abelian variety A by a linear algebraic group L which is a product of a vector group V and a torus T. We obtain therefore an exact sequence

$$0 \longrightarrow V \times T \longrightarrow G \longrightarrow A \longrightarrow 0$$

of algebraic groups which are given a priori as abstract objects. The aim of this section is to furnish in an effective way an explicit description of these group varieties in terms of projective algebraic geometry. We shall concentrate on two main questions. The first concerns the embedding of such varieties in projective space by means of ample invertible sheaves. The second deals with an explicit presentation of the addition morphism of embedded varieties.

As regards the first question, we need first to compactify the group varieties and then to construct some very ample invertible sheaf on the compactification. This problem has been studied by Faltings and Wüstholz [92]. They constructed a certain compactification $\bar{g}$ of the group variety G and a very ample invertible sheaf which gives the desired embedding into projective space. The dimension of the latter was computed by the Riemann–Roch theorem and indeed a general Riemann–Roch formula was proved for the Euler characteristic of line bundles arising naturally in the theory of compactification. This formula leads to Theorem 6.16 below as we shall show; the theorem would seem to be particularly useful in transcendence theory and Diophantine approximation. In fact the dimension of the space of global sections of very ample invertible sheaves appears in the more sophisticated versions of the Siegel lemma and, as we have seen, the latter is a basic ingredient of many of the proofs in this field.

As regards the second question, we denote by $\mu: G \times G \to G$ the addition morphism of G. Then, in connection with the projective embedding described above, μ is given locally by sets of homogeneous polynomials. We need to know the degrees of the polynomials and, in the case when the group is defined over a number field, also information about their heights. In the compact case, that is when the group is an abelian variety, Lange and Ruppert [140] described a method which gives a complete set of addition formulae. Their approach was expanded in Wüstholz [265] to give a description of the addition formulae in general.

We shall now give an account of the main results of [265] which are based on the compactification $\overline{G}$ of G constructed in [92]; here it is assumed that the field of definition is algebraically closed and has characteristic 0. Since G is an extension of an abelian variety by a linear group, it suffices to consider the case when the latter is either the additive or the multiplicative group variety; the results in general follow by taking fibre products. Clearly the particular groups in question can be embedded into $\mathbb{P}^1$ by letting them operate on the projective space through the projective linear group and then taking any dense orbit (cf. Section 4.3). Thus G gets embedded into a fibre space $\overline{q}: \overline{G} \to A$ with compact fibre over the abelian variety. Now there exists on $\overline{G}$ a relatively ample invertible sheaf $\mathcal{O}(1)$ and we begin by determining the Euler characteristic of $\overline{\mathcal{L}} = \mathcal{O}(1) \otimes \overline{q}^*\mathcal{L}$ for a fixed invertible sheaf $\mathcal{L}$ on A. In fact, in the paper [92] referred to above, Faltings and Wüstholz considered the direct image $\mathcal{F} \otimes \mathcal{L}$ of the sheaf $\overline{\mathcal{L}}$, where $\mathcal{F} = \overline{q}_*\mathcal{O}(1)$, and they showed that $\mathcal{F}$ has a filtration

$$\mathcal{F} = \mathcal{F}_0 \supset \mathcal{F}_1 \supset \cdots \supset \mathcal{F}_r \supset \mathcal{F}_{r+1} = 0$$

such that $\mathcal{F}_j/\mathcal{F}_{j+1}$ is in $\mathrm{Pic}^0(A)$ for $j = 0, 1, \ldots, r$. Then, in view of the Riemann–Roch formula discussed in connection with the first question above, we obtain $\chi(\overline{\mathcal{L}}) = r\chi(\mathcal{L})$, where $r = \operatorname{rank} \mathcal{F}$.

Let $\mathbb{K}(\mathcal{L})$ be the theta-group consisting of all $x \in A$ such that $T_x^*\mathcal{L} \cong \mathcal{L}$ where T_x is the translation on A by x; we say that $\mathcal{L}$ is non-degenerate if $\mathbb{K}(\mathcal{L})$ is finite. Now it is proved in [92] that if $\mathcal{L}$ is non-degenerate then there exists a unique $i = i(\mathcal{L})$, with $0 \le i \le \dim A$, such that $H^i(\overline{G}, \overline{\mathcal{L}}) \neq 0$ and $H^k(\overline{G}, \overline{\mathcal{L}}) = 0$ for $k \neq i$. If $\mathcal{L}$ is ample then it is non-degenerate and it is known that $i(\mathcal{L}) = 0$ in this case. Hence we deduce that $H^0(\overline{G}, \overline{\mathcal{L}}) \neq 0$ and furthermore that $\chi(\overline{\mathcal{L}}) = \dim H^0(\overline{G}, \overline{\mathcal{L}})$. From the construction of $\mathcal{F}$, we have $r = 2^d$ where d is the dimension of the linear part of the group G and, by another application of the Riemann–Roch theorem, we obtain $\chi(\mathcal{L}) = \mathcal{L}^n/n!$ where $n = \dim A$. Hence the equation $\chi(\overline{\mathcal{L}}) = r\chi(\mathcal{L})$ stated above gives the following theorem [265].

Theorem 6.16 *If $\mathcal{L}$ is ample then* $\dim H^0(\bar{G}, \overline{\mathcal{L}}) = 2^d \mathcal{L}^n/n!$.

To complete our discussion we return to the question of the degrees and heights of the polynomials occurring in the addition formulae mentioned above. Let G be an extension of an abelian variety A by a commutative linear algebraic group with dimension d. Suppose that $\mathcal{L}_0$ is an ample invertible sheaf on A and put $\mathcal{L} = \mathcal{L}_0^3$. Then $\mathcal{L}$ is very ample and it is known from algebraic geometry that G can be embedded into the projective space $\mathbb{P}(W)$ where $W = H^0(\overline{G}, \overline{\mathcal{L}})$. By virtue of Theorem 6.16, the dimension of W is $N = 2^d \mathcal{L}^n/n!$, and Wüstholz [265] proved further that the addition formulae on G can be expressed completely in terms of sets of bihomogeneous polynomials with bidegree $(3, 3)$.

We remark that the embedding of G into $\mathbb{P}(W)$ described above is not best possible insofar as the dimension of W is concerned. Indeed if one is more careful with the compactification then one can replace W by a vector space with dimension N depending linearly rather than exponentially on d. Moreover Lange and Ruppert [140] suggested that the pair $(3, 3)$ is probably capable of improvement in certain circumstances and in fact Bosma and Lenstra [52] have shown that two addition formulae of bidegree $(2, 2)$ suffice on an elliptic curve.

Finally we mention that the question of giving upper bounds for the heights of polynomials occurring in the addition formulae has not been solved in general but all the techniques for obtaining such estimates seem to be available. The basic tools in this connection are Arakelov theory along with the Geometry of Numbers. To apply Arakelov theory, one would have to develop the theory of Wüstholz [265] in the context of Spec $\mathcal{O}_{\mathbb{K}}$ where $\mathcal{O}_{\mathbb{K}}$ denotes the ring of integers of the number field $\mathbb{K}$ over which the polynomials in question are defined. The main problem seems to be how to obtain effectively Néron models for the relevant groups involved; this is an interesting problem in arithmetical algebraic geometry arising from transcendence theory.

7

The quantitative theory

7.1 Introduction

In the last chapter we established the most natural version of the qualitative theory of logarithmic forms in the context of algebraic groups. Many of the most important applications, however, involve a quantitative form of the theory and this we shall discuss in the present chapter. We shall begin with a report on the results concerning linear forms in ordinary logarithms which refine the basic theory as described in Chapter 2. The estimates given here are fully explicit and they are considerably sharper than those described previously; their derivation depends critically on the theory of multiplicity estimates on group varieties in the form given in Chapter 5. In the following section we report on generalisations to logarithms related to arbitrary commutative algebraic groups. The best general results to date are due to Hirata-Kohno, and more recently Gaudron, and the precision of these is now quite close to those obtainable in the classical case. The work here arises from a long series of earlier researches beginning with publications of Baker and Masser in the elliptic and abelian cases and subsequently taken up especially by Coates, Lang, Philippon and Waldschmidt. This has been a very active area of study and there are, in particular, some significant further contributions in the elliptic case by S. David.

It emerged surprisingly from the early work of Baker on logarithmic forms (see [17]) that if there is a rational linear dependence relation satisfied by logarithms of algebraic numbers then there exists such a relation with coefficients bounded in terms of the heights of the numbers. In Section 7.4 we describe an important body of work that arises from this observation by again considering the general framework of algebraic groups. The results here are due to Masser and Wüstholz and the research was originally motivated by Faltings' famous theorem proving the Mordell conjecture. We shall discuss here how one obtains in this way an effective version of the well-known Tate conjecture which is fundamental to Faltings' paper; in fact we shall describe an isogeny theorem which significantly improves the corresponding result first established by Faltings.

In Section 7.5 we shall give further applications to the arithmetical theory of abelian varieties, in particular to the solution of a problem of Serre on representations of Galois groups, and in Section 7.6 we shall summarise the main implications relating to the Mordell conjecture.

7.2 Sharp estimates for logarithmic forms

We come now to one of the main applications of the theory of multiplicity estimates on group varieties, namely to give a precise lower bound for a linear form in logarithms of algebraic numbers. We apply the work of Section 5.5 which appertains to the group variety $\mathbb{G}_a \times \mathbb{G}_m^r$. This can be interpreted in terms of the zeros of a non-vanishing polynomial P in the ring $\mathbb{C}[Y_0, \ldots, Y_r]$. We introduce differential operators

$$\partial_0 = \frac{\partial}{\partial Y_0}, \quad \partial_j = B_r Y_j \frac{\partial}{\partial Y_j} - B_j Y_r \frac{\partial}{\partial Y_r} \quad (1 \leq j < r),$$

where $B_1, \ldots, B_r$ are rationals with $B_r \neq 0$. We then consider the equations

$$\partial_0^{t_0} \cdots \partial_{r-1}^{t_{r-1}} P\big(s, \vartheta_1^s, \ldots, \vartheta_r^s\big) = 0,$$

where $\vartheta_1, \ldots, \vartheta_r$ are algebraic numbers such that their logarithms are defined and linearly independent over the rationals. Here s is an integer with $0 \leq s \leq \mathcal{S}$ and $t_0, \ldots, t_{r-1}$ are non-negative integers with $t_0 + \cdots + t_{r-1} \leq \mathcal{T}$. Suppose that P has degree at most $\mathcal{D}_j$ in Y_j and that $\mathcal{S}_0 \geq \cdots \geq \mathcal{S}_r \geq 0$ are integers with $\mathcal{S}_0 + \cdots + \mathcal{S}_r \leq \mathcal{S}$. Suppose further that $\mathcal{T}_0 \geq \cdots \geq \mathcal{T}_r \geq 0$ are integers with $\mathcal{T}_0 + \cdots + \mathcal{T}_r \leq \mathcal{T}$. The theory of multiplicity estimates shows that the equations above cannot hold if

$$(\mathcal{S}_m + 1)\binom{\mathcal{T}_m + m + 1 - \delta_{m,r}}{m + 1 - \delta_{m,r}} \geq (m+1)!\,\mathcal{D}_0^{m_0} \cdots \mathcal{D}_r^{m_r}$$

for $m = 0, 1, \ldots, r$ and for all $m_0, \ldots, m_r$, either 0 or 1, where $m_0 + \cdots + m_r = m + 1$ unless there is an algebraic subgroup H of codimension ϱ with $1 \leq \varrho < r$ as described in Theorem 5.7. From (i) or (ii) of Theorem 5.7 we obtain an upper bound for the degree $\delta_j(H)$ of H and Theorem 5.8 gives the estimate $\delta_j(H)$ for the index $\mu_j(H)$. The discussion in Section 5.6 shows how the weighted volume of the lattice associated with H can be expressed in terms of the $\mu_j(H)$ and thus we derive an upper bound for the volume. Now Lemma 4.7 gives a set of primitive linear forms $\mathcal{L}_1, \ldots, \mathcal{L}_\varrho$ in $Z_1, \ldots, Z_r$ with integer coefficients such that

$$\mathcal{L} = B_1 Z_1 + \cdots + B_r Z_r$$

is in the module generated by $\mathcal{L}_1, \ldots, \mathcal{L}_\varrho$ over the rationals. On defining

$$\mathcal{R}_j = \sum_{i=1}^{r} \left(a_i |\partial \mathcal{L}_j / \partial Z_i|\right) \quad (1 \leq j \leq \varrho),$$

where $a_1, \ldots, a_r$ are any positive real numbers, we have either

$$\mathcal{R}_1 \cdots \mathcal{R}_\varrho \mathcal{S}_{\varrho-1} \binom{\mathcal{T}_{\varrho-1} + \varrho - 1}{\varrho - 1} \leq \mathcal{C}(\varrho) \max\left((a_1\mathcal{D}_1)^{m_1} \cdots (a_r\mathcal{D}_r)^{m_r}\right)$$

or the same holds with $\varrho - 1$ on the left replaced by ϱ and $\mathcal{C}(\varrho)$ on the right replaced by $(\varrho+1)\mathcal{C}(\varrho)\mathcal{D}_0$ where the maxima are over all $m_1, \ldots, m_r$ as above with sum ϱ. Here $\mathcal{C}(\varrho)$ is the expression $2^{1-\varrho}(\varrho!)^2(r\varrho)^\varrho$ given by Lemma 4.7 and we are assuming that $\mathcal{D}_j \leq \mathcal{T}_0$ for $1 \leq j \leq r$, that $\mathcal{T}_r < \mathcal{D}_0$ and furthermore that

$$|\mathcal{L}(\log \vartheta_1, \ldots, \log \vartheta_r)| < \pi / \mathcal{S}.$$

Let now $\alpha_1, \ldots, \alpha_n$ be algebraic numbers, not 0 or 1, and let $\log \alpha_1, \ldots, \log \alpha_n$ be fixed determinations of the logarithms. Let $\mathbb{K}$ be the field generated by $\alpha_1, \ldots, \alpha_n$ over the rationals and let d be the degree of $\mathbb{K}$. For each α in $\mathbb{K}$ and any given determination of $\log \alpha$ we define the modified height $h'(\alpha)$ by

$$h'(\alpha) = \frac{1}{d} \max(h(\alpha), |\log \alpha|, 1),$$

where $h(\alpha)$ is the logarithm of the standard Weil height of α defined earlier. In the fundamental paper [33] of Baker and Wüstholz the following result was established concerning the linear form

$$L = b_1 z_1 + \cdots + b_n z_n,$$

where $b_1, \ldots, b_n$ are integers, not all 0.

Theorem 7.1 *If* $\Lambda = L(\log \alpha_1, \ldots, \log \alpha_n) \neq 0$ *then*

$$\log |\Lambda| > -C(n, d)\, h'(\alpha_1) \cdots h'(\alpha_n)\, h'(L),$$

where

$$C(n, d) = 18(n+1)!\, n^{n+1} (32d)^{n+2} \log(2nd).$$

Here we have

$$h'(L) = \frac{1}{d} \max(h(L), 1),$$

where $h(L)$ is the logarithmic Weil height of L, that is $d \log \max(|b_j|/b)$ with b given by the highest common factor of $b_1 \cdots b_n$. The proof of the theorem has its origin in the methods described in the earlier chapters but it incorporates several substantially new aspects. One begins by replacing $\alpha_1, \ldots, \alpha_n$ by a different set $\alpha_1', \ldots, \alpha_r'$ of algebraic numbers such that their logarithms are rational linear combinations of the logarithms of the original set and which satisfy a Kummer condition, that is $\mathbb{K}(\alpha_1'^{1/2}, \ldots, \alpha_r'^{1/2})$ has degree 2^r over the ground field $\mathbb{K}$. The construction combines the classical approach begun in [31] with a technique based on the Geometry of Numbers due to Wüstholz [262].

From the generalised Siegel lemma one obtains a polynomial $P(Y_0, \ldots, Y_r)$ such that

$$\partial_0^{t_0} \cdots \partial_{r-1}^{t_{r-1}} P\big(s, \alpha_1'^{s}, \ldots, \alpha_r'^{s}\big) = 0$$

for integers s with $0 \leq s \leq \mathcal{S}$ and $t_0 + \cdots + t_{r-1} \leq \mathcal{T}$, where $\mathcal{S}$, $\mathcal{T}$ fail to satisfy the multiplicity inequalities indicated above by a small margin. By extrapolation one extends the ranges so that the inequalities become valid for another polynomial arising from the original by Kummer descent. In view of the multiplicity estimate theory mentioned before, it remains then only to eliminate by an inductive procedure the condition regarding the linear forms $\mathcal{L}_1, \ldots, \mathcal{L}_\varrho$. The inductive procedure leads to an auxiliary function in $z_0, \ldots, z_{n-1}$ given by a linear combination over $\lambda_{-1}, \lambda_0, \ldots, \lambda_r$ of expressions of the form

$$\Delta(\gamma_1; t_1) \cdots \Delta(\gamma_{r-1}; t_{r-1})\ \Delta(z_0 + \lambda_{-1}; h, \lambda_0 + 1, t_0)\ e^{\lambda_1 M_1 + \cdots + \lambda_r M_r}$$

where $M_1, \ldots, M_r$ are linear in $z_1, \ldots, z_{n-1}$ and $\gamma_1, \ldots, \gamma_{r-1}$ are linear in $\lambda_1, \ldots, \lambda_r$. In this connection we work with differential operators $\partial_1^*, \ldots, \partial_{n-1}^*$ satisfying

$$\partial_j^* Y_1^{\lambda_1} \cdots Y_r^{\lambda_r} = \gamma_j Y_1^{\lambda_1} \cdots Y_r^{\lambda_r}$$

which generate the same space as $\partial_1, \ldots, \partial_{r-1}$. The extrapolation itself follows on classical lines using the fundamental Cauchy formula but several original features are introduced; in particular, in place of the functions F that occurred earlier one employs finite Blaschke products. This significantly improves upon the numerical precision of the result.

The theorem is capable of generalisation to deal with an arbitrary inhomogeneous linear form

$$\beta_0 + \beta_1 z_1 + \cdots + \beta_n z_n$$

with algebraic coefficients and one would expect essentially the same estimate as given there with $h'(L) + \log(d^n h'(\alpha_1) \cdots h'(\alpha_n))$ in place of $h'(L)$; but the details have yet to be worked out. As already remarked in Section 2.8, Theorem 7.1 has been extended to the p-adic domain in some substantial papers by Kunrui Yu [268] and he obtains a theorem of broadly similar shape. Further, as again discussed in Section 2.8, it is easy to extend the theory so as, when $b_n \neq 0$, to replace $h'(L)$ by

$$\log \max_{1 \leq j < n} \left\{ \frac{|b_n|}{h'(\alpha_j)} + \frac{|b_j|}{h'(\alpha_n)} \right\}.$$

Thus one can show that the analogue of Sharpening II holds, namely, if $b_n = 1$ then

$$\log|\Lambda| > -C(n,d)\max(h'(\alpha_1)\cdots h'(\alpha_n)l^*, \delta B^*)$$

for any $\delta > 0$, where

$$l^* = 2\max(1, \log(\delta^{-1}h'(\alpha_1)\cdots h'(\alpha_{n-1})))$$

and $B^* = \max(|b_j|h'(\alpha_j))$ for $j = 1, 2, \ldots, n-1$. The latter result is of particular interest in connection with the theory of rational approximations to algebraic numbers (see Section 3.3).

As mentioned in Section 2.8 there has been some important new work of Matveev [174] in this context. He has succeeded in replacing the dependence on n in $C(n,d)$ by an expression of the shape c^n for an absolute constant c. The work depends on the construction of a modified auxiliary function where the λ are replaced by numbers forming a free abelian group contained in $\mathbb{Q}^n$ and containing $\mathbb{Z}^n$ and furthermore subject to a double linear restriction, one of which arises from new considerations concerning the Kummer theory. The discussion rests on an estimate for the index of $\mathbb{Z}^n$ in the group referred to above and the best available is currently due to Loher and Masser [148].

7.3 Analogues for algebraic groups

We now discuss briefly the extent to which the classical quantitative theory of logarithmic forms has been carried over to deal with the general situation of commutative group varieties. The most precise work in this field dates back to N. Hirata-Kohno [129] and David [75]. In order to indicate the form of Hirata-Kohno's estimates let $\mathbb{K}$ be a number field and G be a commutative group variety of dimension n defined over $\mathbb{K}$ with Lie algebra $\operatorname{Lie} G$. The complex points on the group G can be considered as a complex Lie group $G(\mathbb{C})$ and there is an associated exponential map

$$\exp_G\colon (\operatorname{Lie} G) \otimes \mathbb{C} \to G(\mathbb{C}).$$

We suppose that G is embedded in some projective space whence we obtain a Weil height on G. On fixing a norm on the complex Lie algebra we then introduce a modified Weil height h' analogous to the modified height that we described in the previous section; now the norm occurs naturally with exponent 2 since we are dealing here with meromorphic functions of order 2 except in the case of a linear group. The basic theory relates to a non-vanishing linear form

$$L(z_0, \ldots, z_n) = \beta_0 z_0 + \cdots + \beta_n z_n$$

with coefficients in $\mathbb{K}$. As before $h'(L)$ will denote the logarithmic Weil height of L. The linear form is evaluated at a point $z_0 = 1$, $z_1 = u_1(v), \ldots, z_n = u_n(v)$ where $u_1, \ldots, u_n$ is a basis for the space dual to the Lie algebra of G and v is an element in the complex Lie algebra such that $\alpha = \exp_G(v)$ is in $G(\mathbb{K})$. We now give a result which can be deduced from the the main theorem in [129] in a straightforward way.

Theorem 7.2 *There exists a positive constant C independent of v and L which can be determined effectively and has the following property. If*

$$\Lambda = L(1, u_1(v), \ldots, u_n(v)) \neq 0$$

then

$$\log|\Lambda| > -C(h'(\alpha))^n (h'(L) + h'(\alpha))(\max(1, \log(h'(L)h'(\alpha)))^{n+1}.$$

Clearly this is an inhomogeneous analogue of Theorem 7.1 which deals with the case when G is a product of multiplicative groups. On comparing results one sees that Theorem 7.2 is of essentially the same shape as that obtainable for ordinary logarithms except for the second order term appearing to the power $n+1$. The main novelty in the proof is the introduction of a further auxiliary additive group in the definition of G and this leads to a substantial improvement on previous results in the field which involved powers of $h'(L)$ depending on n. The principal problem in this area until recently was to eliminate the second order term involving $h'(L)$ from the estimate in Theorem 7.2. This was solved by Gaudron [101] (see also David and Hirata-Kohno [76] for an earlier announcement in the elliptic case) using some new ideas involving a change of variables method which he attributes to G. V. Chudnovsky. Thus he showed that $\log|\Lambda| > -c\,h'(L)$ for some $c = c(\alpha) > 0$; the estimate is best possible in terms of the height of L and it is the precise analogue of the classical logarithmic form inequality given by Theorem 2.5.

The constant C in Theorem 7.2 has a much more complicated structure than the expression $C(n,d)$ occurring in the last section. One reason is that the group itself depends on parameters related to moduli spaces which must naturally appear in C. Also the projective embedding depends on parameters associated with Picard groups and these too have to appear in C. Finally we made a choice of the metric on the complex Lie algebra of G and this again will be reflected in the constant. As far as we are aware the only detail relating to C that so far has been worked out explicitly is the dependence on the degree d of $\mathbb{K}$ and this appears as a power of d just as in Theorem 7.1.

In the special case when G is a product of elliptic curves, a fully explicit expression for C has been worked out in David's paper [75] cited above. The result has been applied to yield an alternative method for finding all the integer points on a given elliptic curve in Weierstrass form defined over the field of rational numbers. The approach is in some respects more direct than the original technique as described in Chaper 3 which involved reduction to S-unit equations and linear forms in ordinary logarithms but it requires the knowledge of an explicit set of generators for the basis of the Mordell–Weil group. In general, as is well known, the Mordell–Weil theorem is ineffective in this respect but some efficient algorithms are known in practice which usually yield the desired computation.

The elliptic logarithm method depends on the fact that the Weil height $h(P)$ of an integral point P varies as $-\delta(P,0)$ where δ denotes a logarithmic distance function and 0 is the neutral element on the curve. This follows for instance from elementary properties of the Weierstrass function which furnishes the exponential map. By the Mordell–Weil theorem we can write

$$P = m_1 P_1 + \cdots + m_r P_r + Q,$$

where $P_1, \ldots, P_r$ is a basis for the group of rational points and Q is in the torsion subgroup. On taking elliptic logarithms the equation becomes

$$v = m_1 v_1 + \cdots + m_r v_r + (m/q)\,\omega + (m'/q)\,\omega'$$

for basis elements ω, ω' of the period lattice and integers m, m', q with q dividing the order of the torsion group. Now, as is well known from the theory of the Néron–Tate height, the maximum B of the absolute values of the m satisfies $B \ll (h(P))^{1/2}$, assuming that $v, v_1, \ldots, v_r$ are chosen in the fundamental domain. On applying Theorem 7.2 in the elliptic case we get a lower bound for $\log |v|$ that varies as $-(\log B)(\log\log B)^{\lambda}$ where λ depends linearly on r. We fix the function δ, as we may, so that $\delta(P, 0) = \log |v|$, and then it follows that $h(P) \ll (\log h(P))(\log\log h(P))^{\lambda}$. This gives a bound for $h(P)$ depending on, amongst other things, the heights of a set of generators for the Mordell–Weil group. Hence, in the cases when one is able to find a suitable set of generators, an appeal to S. David's explicit expression for C, together with computational techniques of the kind described in Section 3.5, shows that all integer points P on the curve can be determined effectively.

The method just described was first successfully applied by Stroeker and Tzanakis [240] and, at about the same time, by Gebel, Pethő and Zimmer [102]. For instance, in [240], the authors gave the complete set of solutions in integers x, y of the equation

$$y^2 = (x + 337)(x^2 + 337^2);$$

in this case it turns out that the rank of the Mordell–Weil group is 3 and a complete set of generators can be determined explicitly. Another example, from [102], is the equation

$$y^2 = x^3 - 1642032x + 628747920$$

where there is a basis consisting of six integer points. The elliptic logarithm method has been widely developed and there is a good discussion of its ramifications in the survey article of Győry [123]. In particular, as mentioned there, the condition that the elliptic curve has Weierstrass form can be relaxed, the work can be extended to number fields and to S-integral points and moreover to certain quartic equations; for further details see the book by Smart [232].

We remark that there is a close connection here with Siegel's famous theorem on the finiteness of the number of integer points on curves of genus at least 1. Siegel too appealed to the Mordell–Weil theorem but his argument took a different path to that described above, involving coverings and the Thue–Siegel theorem. Now that we have a theory of linear forms in abelian logarithms, a more direct proof of Siegel's theorem can be given following the above approach. It shows, in particular, that an effective Mordell–Weil theorem would imply an effective Siegel theorem. It is of interest to note that there was already some indication of this train of argument in a paper of Lang [135].

Finally we mention that there is an extensive literature on analogues of logarithmic form results in the context of Drinfeld modules; for basic references see the papers by Brownawell [60], by Jing Yu [266] and the book by Thakur [242]. In particular Bosser [53] gives analogues for Drinfeld logarithms of the theorems of Hirata-Kohno and David referred to at the beginning of this section and there is definitive work on algebraic relations among the values of the Γ-function by Anderson, Brownawell and Papanikolas [5].

7.4 Isogeny theorems

In the past two decades transcendence theory has developed very powerful tools in order to deal with Diophantine questions. One striking example was Shafarevich's theorem on the finiteness of the number of isogeny classes of elliptic curves over a number field with good reduction outside a finite set of finite places. Here the problem of bounding this number was reduced to the famous Mordell equation $y^2 = x^3 + k$ for which effective upper bounds for the height of the solutions in terms of k were given using the theory of linear forms in logarithms (see Section 3.3). Another example is the proof of Tate's conjecture in the special case of elliptic curves over the field of rational numbers given by D. V. and G. V. Chudnovsky [65]. An entirely new method was found by Masser and Wüstholz [168] leading to a proof of Tate's conjecture for elliptic curves in general; the approach is based on linear forms in elliptic logarithms. It gives an estimate for isogenies between elliptic curves from which Shafarevich's theorem follows readily. In order to state the result, let E be the elliptic curve with equation

$$y^2 = 4x^3 - g_2 x - g_3,$$

where g_2, g_3 are elements of an algebraic number field $\mathbb{K}$ with degree d. We define the height of E by

$$h(E) = \max(1, h(g_2), h(g_3)),$$

where $h(g)$ denotes the absolute logarithmic Weil height of g.

Theorem 7.3 (Elliptic isogeny theorem) *There exists an effectively computable positive number $C = C(d)$ such that if E' is an elliptic curve over $\mathbb{K}$ isogenous to E then there is an isogeny φ between E and E' with*

$$\deg \varphi \leq C(h(E))^4.$$

Here E and E' are said to be isogenous if there is a homomorphism φ from E to E' with finite kernel; the degree of the isogeny φ is the number of elements in the kernel. A notable feature of the theorem is the fact that C depends on the degree rather than on the discriminant of $\mathbb{K}$ as was initially expected by some experts in the field. An improvement in the result, replacing 4 by 2 in the exponent, was obtained by Pellarin [190] and here an estimate for C was calculated explicitly. As an application of Theorem 7.3, Masser and Wüstholz [167] obtained the effective bound $C(h(E))^8$ for the number of $\mathbb{K}$-isomorphism classes of elliptic curves that are defined over $\mathbb{K}$ and are $\mathbb{K}$-isomorphic to E.

The idea of the proof is very simple to explain. Let Ω and Ω' be the period lattices for E and E' respectively. Then an isogeny φ induces a homomorphism φ_* from Ω to Ω'. Hence one obtains an algebraic number $\alpha = \alpha(\varphi)$ with

$$\alpha\Omega \subseteq \Omega'.$$

This gives a set of dependence relations

$$\alpha\omega_i = m_{i1}\omega_1' + m_{i2}\omega_2' \quad (i = 1, 2),$$

where ω_1, ω_2 and ω_1', ω_2' are a basis for Ω and Ω' respectively. By transcendence methods based on the observation mentioned in Section 7.1 one obtains new relations with bounds for the m_{ij} depending only on d and $h(E)$. The relations give then a new isogeny φ with bounds on the degree as stated in the theorem. The basic structure of the argument generalises to higher dimensional abelian varieties but many technical problems arise. Masser and Wüstholz [170, 171] succeeded in overcoming these and thus established the desired abelian analogue of Theorem 7.3. It is given by Theorem 7.5 below.

The main ingredient in the proof of the latter is a quantitative version of the analytic subgroup theorem. Let A be an abelian variety over $\mathbb{K}$ with dimension g and with Faltings height $h_F(A)$ (see [94]). Further, let L be a positive element in $(\operatorname{Pic} A)(\mathbb{K})$ so that L is a polarisation and determines an Hermitian form H on the complex vector space $V = \operatorname{Lie} \otimes_{\mathbb{K}} \mathbb{C}$. Then the complex points $A(\mathbb{C})$ of A form a complex manifold isomorphic to V/Λ for some lattice $\Lambda \subseteq V$. By construction, the imaginary part E of H is integer valued on Λ and non-degenerate. We choose a symplectic basis $\gamma_1, \ldots, \gamma_g, \delta_1, \ldots, \delta_g$ for Λ and this determines a point $\tau = (\tau_{ij})$ in the Siegel upper half-plane $\mathbb{H}_g$ by the equations

$$\gamma_j = \sum_{i=1}^{g} \tau_{ij}\, \delta_i \quad (1 \le j \le g).$$

Now the matrix $\varepsilon = \big(E(\gamma_i, \delta_j)\big)$ is diagonal and the period matrix for A is given by the $g \times 2g$ matrix (ε, τ). The degree d_L of the polarisation L is defined as the square of the determinant of ε. For $\omega \in \Lambda$ we take B_ω to be the smallest abelian subvariety B such that $\omega \in (\operatorname{Lie} B) \otimes_{\mathbb{K}'} \mathbb{C}$ where $\mathbb{K}'$ is the field of definition of B. Then we have the following result.

Theorem 7.4 (Quantitative analytic subgroup theorem) *There exist effectively computable numbers C and c such that*

$$\deg B_\omega \le C\big(h'_F(A) + H(\omega, \omega)\big)^c.$$

Here $h'_F(A)$ is the modified Faltings height given by $\max(1, h_F(A))$*; further C depends only on g, d and d_L, and c depends only on g.*

We have stated the theorem in terms of the Faltings height $h_F(A)$ since this is an intrinsic quantity attached to an abelian variety. However, the result can be expressed alternatively in terms of the theta-height $h_L(A)$ which is a direct analogue of the height of an elliptic curve defined earlier. The quantities can be related by way of the theory of moduli spaces and it turns out that $h_L(A)$ is equal to $d_L h_F(A)$ up to a second order term.

Further, by the Geometry of Numbers, one can find a basis $\omega_1, \ldots, \omega_{2g}$ for the period lattice Λ such that if $\omega = \omega_j$ then $H(\omega, \omega)$ is bounded in terms of $h'_F(A)$; for such a basis element, the conclusion of the theorem simply takes the form $\deg B_\omega \leq C\big(h'_F(A)\big)^c$ and this is the result we shall use subsequently. We remark also that the theorem is almost certainly capable of generalisation to an unrestricted commutative group variety and an arbitrary logarithm of an algebraic point rather than just a period vector as here; however, the details remain to be worked out. The proof of Theorem 7.4 runs on the same lines as that of the analytic subgroup theorem but it involves, necessarily, the development of a theory of Siegel modular forms and theta-functions for this particular context; this was accomplished in the paper of Masser and Wüstholz [170]. The dependence of C on the dimension g in the latter work was not clearly effective but Bost [54] made it evident that an explicit expression could in fact be obtained.

Now let A be an abelian variety with dimension g and with a polarisation L as above. It is shown in [171] that Theorem 7.4 yields the following result.

Theorem 7.5 (Isogeny theorem) *There exist effectively computable positive numbers C and c such that if A' is an abelian variety over $\mathbb{K}$ isogenous to A and if L' is a polarisation on A' then there is an isogeny φ from A to A' with*

$$\deg \varphi \leq C\big(h'_F(A)\big)^c,$$

where C depends only on g, d, d_L and $d_{L'}$, and c depends only on g.

As in the elliptic case, we say that abelian varieties A and A' are isogenous if there is a homomorphism φ from A to A' with finite kernel; the degree of φ is defined as the cardinality of the kernel. Note that the definition here does not involve any reference to polarisations. In fact it seems almost certain that the theorem can be expressed completely independently of L and L'. This would follow immediately if the degree of some polarisation on an abelian variety A could be estimated from above in terms of $h'_F(A)$ and such a bound has already been obtained in most cases; the problem is connected with Albert's classification of algebras with positive involution.

For the proof of Theorem 7.5 one begins by choosing a minimal isogeny ψ from A to A' and, as in the case of elliptic curves, one takes period lattices Λ and Λ' and obtains a set of dependence relations between the periods of A and the periods of A'. More precisely, if $\omega_1, \ldots, \omega_{2g}$ and $\omega'_1, \ldots, \omega'_{2g}$ denote bases for Λ and Λ' then for each $\omega = \omega_j$ one gets an equation

$$\varphi_*(\omega) = m_1\omega'_1 + \cdots + m_{2g}\omega'_{2g},$$

where φ_* is the tangent map and $m_1, \ldots, m_{2g}$ are integers. Thus the vector $\eta = (\omega, \omega_1, \ldots, \omega_{2g})$ is non-zero and is contained in a subspace of $\operatorname{Lie} G$ with $G = A \times A'^{2g}$; the subspace is defined by equations of the form

$$\varphi_*(v) = m_1 v'_1 + \cdots + m_{2g} v'_{2g},$$

where v is in $\operatorname{Lie} A$ and $v'_1, \ldots, v'_{2g}$ belong to $\operatorname{Lie} A'$. By the version of Theorem 7.4 discussed above we have the estimate $\deg B_\eta \leq C\big(h'_F(G)\big)^c$. Now, by the additive property of the Faltings height, $h'_F(G)$ is bounded in terms of $h'_F(A)$. Further, one can construct from B_η the graph of an isogeny from B' to B where B and B' are abelian subvarieties of A and A' respectively. Furthermore, the estimate above implies that the degree of the isogeny and the degrees of the polarisations induced on B and B' are all bounded in terms of $\deg B_\eta$. Theorem 7.5 follows in the case when A is simple and the general conclusion can be reduced to this by induction (see [171]).

7.5 Discriminants, polarisations and Galois groups

In this section we shall discuss three applications of Theorems 7.4 and 7.5. Firstly we shall give an estimate for the discriminant of the ring of endomorphisms of an abelian variety associated with some fixed polarisation. Secondly we shall use the result to establish the existence of a polarisation with degree bounded in terms of a power of the Faltings height. Finally we shall discuss an application of the isogeny theorem to Galois representations attached to an elliptic curve and the connection with a problem of Serre.

Let A be an abelian variety over a number field $\mathbb{K}$ of degree d, let g be the dimension of A and let L be a polarisation on A. In analogy with the usual discriminant of a number field, Masser and Wüstholz [172] introduce a discriminant $\mathcal{D}_L > 0$ for the ring of endomorphisms $\operatorname{End} A$ of A using the Rosati involution and the reduced trace (see [184]). They then establish the following result.

Theorem 7.6 (Discriminant theorem) *There exist effectively computable positive numbers C and c such that*

$$\mathcal{D}_L \le C\big(h'_F(A)\big)^c,$$

where C depends only on g, d and d_L, and c depends only on g.

For the proof, one assumes first that A is a simple abelian variety and one chooses an endomorphism of A. Then, as in Theorem 7.5, this gives a set of period relations which is utilised to define a linear subspace of $\operatorname{Lie} G$ with $G = A^{2g+1}$. By Theorem 7.4, there is an abelian subvariety of G with degree bounded in terms of $h_F(A)$; the existence of this subvariety enables us to construct endomorphisms of A which, by varying the period ω, generate a sublattice of $\operatorname{End} A$. Now the construction furnishes a bound for the degrees of these generators in terms of $h_F(A)$ and thus a similar bound for the discriminant of the sublattice. Since the discriminant of a lattice does not exceed that of a sublattice, the theorem follows at once for simple abelian varieties; the general result is then obtained by induction.

The bound for the discriminant $\mathcal{D}_L$ in Theorem 7.6 depends on d_L and an obvious question arises as to whether there exists a polarisation L defined over $\mathbb{K}$ whose degree d_L can be bounded in terms of d, g and $h_F(A)$. A positive answer to this question in the simplest case is given by the following theorem established by Masser and Wüstholz [173].

Theorem 7.7 (Polarisation theorem) *Suppose that* $\operatorname{End} A = \mathbb{Z}$. *Then there exist effectively computable positive numbers C and c and a polarisation L of A with degree d_L satisfying*

$$d_L \le C\big(h'_F(A)\big)^c,$$

where C depends only on g and d, and c depends only on g.

Plainly, in view of Theorem 7.7, the number C in Theorem 7.5 in the case when $\mathrm{End}\,A = \mathbb{Z}$ can be determined in terms of g and d only. The proof of Theorem 7.7 is an application of Theorem 7.5 but there are several novel ingredients; in particular, the Geometry of Numbers plays an important role. In later work, as yet unpublished, Masser and Wüstholz deal with more general endomorphism rings and here the theory of involution algebras is utilised. It remains an open question as to whether Theorem 7.7 holds for arbitrary abelian varieties defined over an algebraic number field and moreover whether a bound for d_L exists that does not involve $h_F(A)$.

Finally we remark that the isogeny theorem has been applied to a problem of Serre concerning Galois representations attached to elliptic curves. Let $\mathbb{K}$ be a number field, let $\overline{\mathbb{K}}$ be its algebraic closure and let $\pi = \mathrm{Gal}(\overline{\mathbb{K}}/\mathbb{K})$. If E is an elliptic curve defined over $\mathbb{K}$ then π acts on the $\overline{\mathbb{K}}$-rational points $E(\overline{\mathbb{K}})$ of E and, in particular, on the group E_ℓ of points of order dividing ℓ. When ℓ is a prime number, E_ℓ is a vector space over the finite field $\mathbb{F}_\ell = \mathbb{Z}/\ell\mathbb{Z}$ and, by the action of π, we obtain a representation $\varrho_\ell\colon \pi \to \mathrm{GL}(E_\ell)$. In the case when E does not have complex multiplication over $\overline{\mathbb{K}}$, a fundamental result of Serre [217, 218] asserts that there exists a constant $\ell_0 > 0$ such that $\varrho_\ell(\pi) = \mathrm{GL}(E_\ell)$ for all $\ell > \ell_0$. The proof depends on the theory of ℓ-adic representations and is ineffective, that is, it does not enable a general estimate for ℓ_0 to be written down. Serre gave a number of effective examples and results for special classes of elliptic curves; for instance, he obtained in [218] a simple estimate for ℓ_0 when $\mathbb{K} = \mathbb{Q}$ and E is semistable and later, in [219], he removed the semistability condition by assuming the generalised Riemann hypothesis. Masser and Wüstholz [169] succeeded in giving the first demonstration leading to a general and effective estimate for ℓ_0. Their result followed fairly readily from Theorem 7.3 together with a consideration of isogenies of two-dimensional abelian varieties and the group theoretical analysis of Serre [218]. In the case when ℓ does not divide the discriminant of $\mathbb{K}$, the lower bound ℓ_0 for ℓ takes the form $C(\max(d,h))^c$, where h is the absolute logarithmic Weil height of the j-invariant of E (so that the result holds with $h = h(E)$ when E has the Weierstrass equation as in Section 7.4) and C, c are absolute positive constants.

7.6 The Mordell and Tate conjectures

Mordell conjectured that any curve of genus at least 2 defined over the rationals has only a finite number of rational points. This contrasts with the cases of genus 0 and 1 where it is classically known that there can be infinitely many such points. Indeed Mordell [181] himself established a famous result in this context to the effect that the group of rational points on a curve of genus 1 is finitely generated. The problem became one of the fundamental goals in Diophantine geometry. After earlier work of Manin and Grauert in the function field case, Faltings [90] succeeded in 1983 in establishing Mordell's conjecture and, in fact, his work applied to any curve of genus at least 2 defined over an algebraic number field. Faltings' argument involved deep results from arithmetic algebraic geometry including the theory of moduli spaces and the representation theory of finite group schemes going back to Raynaud. Moreover, a crucial step in the proof was the verification of a celebrated conjecture of Tate [241] on modules over the l-adic numbers associated with an abelian variety (see e.g. Lang [138]).

The isogeny theorem discussed above leads to a new and more direct proof of Faltings' theorem. In fact it gives an immediate verification of Tate's conjecture quite different from that originally discovered by Faltings. The connection with isogenies used here was apparently first described by Lichtenbaum; see the discussion in Tate [241]. Indeed an assertion closely related to Theorem 7.5 appeared as a hypothesis in [241] long before Masser and Wüstholz proved their theorem.

The isogeny theorem also allows one to simplify Faltings' original train of argument linking Tate's conjecture with the Mordell conjecture; this rested on a construction of Kodaira and Parshin. Principally, the theorem enables one to show that an isogeny class contains only finitely many isomorphism classes of abelian varieties with good reduction outside a finite set of places. The latter eliminates much of the sophistication associated with moduli spaces and the Faltings height occurring in [90]. Furthermore, the Masser–Wüstholz argument furnishes a new proof of a conjecture originally formulated by Shafarevich on the finiteness of the total number of all such isomorphism classes. Shafarevich himself outlined a proof in the case of elliptic curves and the conjecture was confirmed in general as a component of Faltings' work; in fact Faltings showed that Shafarevich's conjecture follows as a consequence of Tate's conjecture. However, neither Shafarevich's approach in the elliptic case nor that of Faltings in general is effective and it is only as a consequence of the transcendence method of proof of Theorem 7.5 that one can now give an effective finiteness theorem (see [171]).

It seems likely that this area of study will prove capable of much wider geometrical interpretation as has already been indicated in the article of Bost [54].

8
Further aspects of Diophantine geometry

8.1 Introduction

There is a clear division in mathematics between results that one terms 'effective' and others that one terms 'ineffective'. Most of the material discussed so far in this book relates to the theory of logarithmic forms and this is certainly effective. The typical case is an application to a Diophantine equation and here the theory leads in general to an explicit bound for all the integer variables, whence the complete solution can be determined in principle (and often in practice) by a finite amount of computation. There is however a substantial body of work in Diophantine geometry and elsewhere which enables one to decide whether there are finitely or infinitely many solutions to a particular problem, and indeed in general to give an estimate for the number of solutions when finite, but which cannot in principle lead to a complete determination. It is consequently termed 'ineffective'. Some famous instances here are the Schmidt subspace theorem, Faltings' theorem on the Mordell conjecture and Siegel's theorem on the class number of imaginary quadratic fields. Historically these have been important in the overall evolution of mathematics and to conclude our book we give a short discussion of some of these topics.

8.2 The Schmidt subspace theorem

In 1972 W. M. Schmidt [209] (see also [210]) obtained a far-reaching generalisation of the Thue–Siegel–Roth theorem, that is Theorem 1.7, relating to simultaneous Diophantine approximation. Let $L_1, \ldots, L_n$ be

$n \geq 2$ linearly independent linear forms in $x_1, \ldots, x_n$ with algebraic coefficients. Suppose that $\delta > 0$ and consider the inequality

$$|L_1(\mathbf{x}) \cdots L_n(\mathbf{x})| < |\mathbf{x}|^{-\delta},$$

where $\mathbf{x} = (x_1, \ldots, x_n)$ and $|\mathbf{x}| = (x_1^2 + \cdots + x_n^2)^{1/2}$. Then Schmidt proved the following.

Theorem 8.1 *There exists a finite set $T_1, \ldots, T_t$ of proper linear subspaces of $\mathbb{Q}^n$ such that all solutions of the above inequality in integers $x_1, \ldots, x_n$ are contained in the union of $T_1, \ldots, T_t$.*

The particular case when $L_1 = x_2 - \alpha x_1$, $L_2 = x_1$ shows at once that the inequality $|L_1 L_2| < \max(|x_1|, |x_2|)^{-\delta}$ has only finitely many solutions in non-zero integers x_1, x_2 unless α is rational; this is just Theorem 1.7. More generally, on taking $\alpha_1, \ldots, \alpha_m$ as algebraic numbers and

$$L_1 = x_n - \alpha_1 x_1 - \cdots - \alpha_m x_m, \qquad L_2 = x_1, \ldots, L_n = x_m,$$

where $n = m + 1$, we deduce that the inequality

$$|x_1 \cdots x_m|^{1+\delta} \|x_1 \alpha_1 + \cdots + x_m \alpha_m\| < 1$$

has only finitely many solutions in non-zero integers $x_1, \ldots, x_m$ unless 1, $\alpha_1, \ldots, \alpha_m$ are linearly dependent over the rationals; here $\|x\|$ denotes the distance of x to the nearest integer, taken positively. This is the basic Schmidt result on simultaneous rational approximation to algebraic numbers; it is discussed and a proof is given in [25, Ch. 7].

The main applications of Theorem 8.1 together with its natural generalisations to the p-adic domain have been to studies on norm form equations, on unit equations and on linear recurrence sequences. As a particular instance we mention the work of Evertse, Schlickewei and Schmidt [89] which gives a bound for the number of solutions of the equation

$$a_1 x_1 + \cdots + a_n x_n = 1$$

in which it is assumed that no proper subsum on the left vanishes; here the elements $x_1, \ldots, x_n$ lie in a subgroup of $(\mathbb{K}^*)^n$ of finite rank where $\mathbb{K}^*$ is the multiplicative group of non-zero elements of an algebraically closed field $\mathbb{K}$ of characteristic zero. For extensive discussion of this area of work see the papers of Schlickewei and Schmidt in [3, pp. 107–247].

The proof of Theorem 8.1 as given originally by Schmidt was based on the classical work of Thue, Siegel, Schneider, Gelfond, Dyson and Roth as referred to in Section 1.1. They considered a single algebraic irrationality α and studied the question as to the real exponents $\varkappa$ for which there exist only finitely many coprime integers p, q with $q > 0$ and with

$$|\alpha - p/q| < q^{-\varkappa}.$$

The method that they developed consisted in choosing a sequence of pairs p_n, q_n satisfying the inequality and such that

$$\log q_{n+1} \gg \log q_n$$

with a sufficiently large implied constant on the right. This can be done if one assumes that the inequality has infinitely many solutions p, q. One fixes m such solutions for some sufficiently large integer m and one then chooses positive integers $d_1, \ldots, d_m$ such that

$$d_j \log q_j \approx d_1 \log q_1 .$$

This enables a polynomial $P(x_1, \ldots, x_m)$ to be constructed with integer coefficients and with degree d_j in x_j such that P vanishes at $(\alpha, \ldots, \alpha)$ to a high weighted order, termed the index of P. The derivations $\partial/\partial x_1, \ldots, \partial/\partial x_m$ have different weights $1/d_1, \ldots, 1/d_m$ and the multiplicity is taken with weights. Using the assumed approximations $p_1/q_1, \ldots, p_m/q_m$ one shows that P continues to have a positive index at the point $(p_1/q_1, \ldots, p_m/q_m)$. To complete the proof of Theorem 1.7 one has to demonstrate that the latter cannot happen and this is the content of the classical Roth lemma utilising generalised Wronskians; an alternative approach based on Dyson [81] was later given by Vojta [249]. Schmidt followed the same approach but utilised polynomials P in kn variables x_{lm} $(1 \leq l \leq k,\ 1 \leq m \leq n)$, homogeneous in $x_{1m}, \ldots, x_{km}$ for each m. The index is now defined in terms of n-fold products of subspaces $T_1, \ldots, T_n$ of $\mathbb{Q}^k$ and extensive use is made of the Geometry of Numbers, in particular, the theory of successive minima. We refer to [3, pp. 107–170] and also to [48] for detailed discussions.

8.3 Faltings' product theorem

A new proof of Theorem 8.1, substantially different from that of Schmidt, was given by Faltings and Wüstholz [93]. In particular, they replaced the Roth lemma by a deep generalisation, termed the Faltings product theorem [91]. This is a result about the vanishing of multihomogeneous polynomials on products of projective spaces and it has its origins in work of Vojta [249] in 1987.

Vojta succeeded in making the Thue–Siegel–Roth method work in a more geometrical situation by replacing the underlying space $\mathbb{P}^1$ with an arbitrary curve of genus $g > 1$. The existence of infinitely many rational points on such a curve leads to certain Diophantine inequalities analogous to the inequality for $\alpha - p/q$ above and, by way of Arakelov theory and an arithmetic Riemann–Roch theorem, he constructed a section of a certain line bundle on the product of the curve with itself. The section replaces the polynomial P from above and it vanishes to a certain weighted order; Vojta showed that this cannot happen by utilising a version of the Dyson lemma due to Viola.

Shortly afterwards Faltings generalised Vojta's geometric method in an ingenious way. He replaced the product of two copies of a single curve by a product of any number of copies of an abelian variety and thus he was able to prove a conjecture of Weil and Lang to the effect that a subvariety of an abelian variety defined over a number field $\mathbb{K}$ contains only finitely many $\mathbb{K}$-rational points unless it is a translate of a positive dimensional abelian subvariety. This result contains the Mordell conjecture discussed in Section 7.6; indeed any curve of genus >1 can be embedded as a subvariety into its Jacobian and, since the genus condition implies that the curve cannot be a translate as above, it follows that there are only finitely many rational points. The same techniques enabled Faltings to prove a Diophantine approximation theorem on abelian varieties and thus he deduced another conjecture of Lang, namely that Siegel's result on integral points on curves (see Section 3.4) remains true for an affine open subset of an abelian variety. Moreover, Vojta [251] succeeded in extending everything to semi-abelian varieties, that is extensions of abelian varieties by tori. One of the main problems in this context was to find a general replacement of the Dyson lemma in order to bound the weighted multiplicity of a section of a multihomogeneous line bundle, that is, the so-called Faltings bundle. Faltings succeeded in overcoming the problem by establishing his product theorem.

It turns out that there is a close connection between the product theorem and the theory of multiplicity estimates on group varieties as discussed in earlier chapters. Thus, for instance, there is an effective version of the product theorem due to Ferretti [99] and the proof here uses arithmetic intersection theory, in particular a kind of arithmetic Bezout theorem, as well as results of Masser and Wüstholz [166] concerning an effective ideal theory and a result of Wüstholz [263] relating to the computation of multiplicities of the primary components of ideals.

8.4 The André–Oort conjecture

Faltings' theorem referred to above which verified the Weil–Lang conjecture on rational points on subvarieties of abelian varieties can be seen as a basic result relating number theory and geometry. It shows indeed that an arbitrary set S of $\mathbb{K}$-rational points on an abelian variety over $\mathbb{K}$ determines the geometric structure of its Zariski closure $\overline{S}$. For clearly $\overline{S}$ is a finite union of subvarieties defined over $\mathbb{K}$ and, by Faltings' theorem, the components are all translates of abelian subvarieties by some point of S.

This intimate relation between number theory and geometry appears again in the context of Shimura varieties. They were introduced in order to study moduli problems for abelian varieties with prescribed endomorphism algebras and the modern theories originate from work of Deligne in the 1970s. The simplest example is the moduli space of elliptic curves typically denoted by $X_0(1)$; it is given by the quotient of the upper half-plane by the modular group $\mathrm{SL}_2(\mathbb{Z})$ and the classical j-function gives an isomorphism of $X_0(1)$ to the affine line. Other examples are the Siegel and Hilbert modular varieties typically denoted by $\mathcal{A}_g$ and $\mathcal{H}_g$; these classify respectively abelian varieties of dimension g with a principal polarisation and with an endomorphism algebra containing a totally real number field. Deligne showed that every Shimura variety can be defined over a unique number field which is termed the reflex field of the variety. Further, he introduced the notion of 'special points'; these form a dense subset of the $\overline{\mathbb{Q}}$-rational points on the variety and each special point corresponds to a Shimura subvariety of dimension 0. The special points in the case of $X_0(1)$ parametrise the elliptic curves with complex multiplication and are represented in the upper half-plane by imaginary quadratic numbers. Similarly one can characterise special points on $\mathcal{A}_g$ and $\mathcal{H}_g$; they correspond to abelian varieties with complex multiplication and, in the case of $\mathcal{H}_g$, they are the so-called Heegner points. Deligne showed that the set of special points on a Shimura variety is dense with respect to the complex topology. Clearly it is invariant under the action of the absolute Galois group of the reflex field and moreover, as is easily seen, it is also invariant under the action of the Hecke algebra of the group associated with the Shimura variety.

In analogy with the observation at the beginning of this section about the geometric structure of the Zariski closure $\overline{S}$ of the set S indicated there, we now consider a set S of special points on a Shimura variety. André in 1989 and, independently, Oort in 1994 conjectured that the irreducible components of $\overline{S}$ are Shimura subvarieties of Hodge type; this means that they are irreducible components of images of Shimura subvarieties by Hecke operators. The first result in support of the conjecture was obtained by André [6] in 1998; he verified the simplest non-trivial case of the assertion namely when the Shimura variety is the moduli space $X_0(1) \times X_0(1)$ of products of two elliptic curves. In this instance the function $j \times j$ gives an isomorphism to the affine plane and the Shimura subvarieties comprise the modular curves $X_0(N)$ for integers $N \geq 1$ in conjunction with the horizontal and vertical axes. Here $X_0(N)$ is defined by the classical modular equation relating $j(z)$ and $j(Nz)$ and a point on the curve corresponds to a pair of elliptic curves together with a cyclic isogeny of degree N. André's proof rested on classical class field theory and he made a critical appeal to an approximation result of Masser concerning numbers τ such that $j(\tau)$ is algebraic to complete his argument. Independently and at around the same time, Edixhoven [82] gave a totally different demonstration of the result by way of the theory of modular curves and Hecke operators but he assumed here the validity of the generalised Riemann hypothesis for imaginary quadratic fields.

Subsequently André and Edixhoven extended their proofs to Hilbert modular surfaces and more recently Edixhoven and Yafaev [83] have succeeded in proving the André–Oort conjecture in the case when S is contained in the Hecke orbit of a single special point. The latter study was motivated by a paper [257] of Wolfart on algebraic values of hypergeometric functions; it turned out, as observed by Gubler, that the proof of Wolfart's theorem in [257] was not correct but Cohen and Wüstholz [74] showed that one could restore the result if one could verify a weak form of the André–Oort conjecture and this was proved in the work of Edixhoven and Yafaev. We shall discuss the transcendence background to the hypergeometric theory in the next section.

Finally we remark that the André–Oort conjecture remains an object of intensive study and there is valuable new work in this context by Yafaev and Klingler. It involves the equidistribution theory on semi-abelian varieties that originated during the last decade from researches of Szpiro, Zhang and others and the extension of this theory to Hecke orbits on connected linear algebraic groups by Clozel, Oh and Ullmo. The latter has been applied by Pink in the context of mixed Shimura varieties to throw interesting light on attempts to unify all the conjectures of Mordell and André–Oort type.

8.5 Hypergeometric functions

Transcendental functions can be expected in general to take transcendental values at algebraic values of the argument. But there are exceptions. The simplest example is given by the transcendental function $e^{i\pi z}$ which, for rational z, is a root of unity. Nonetheless, by the Gelfond–Schneider theorem (see Section 2.2), if z is an algebraic irrational, then $e^{i\pi z}(= (-1)^z)$ is indeed transcendental. In other words, the rational numbers can be characterised as the complex numbers z for which the functions z and $e^{i\pi z}$ are both algebraic. Another example is given by the classical j-function. As mentioned in Section 2.3, it follows from a theorem of Schneider that the imaginary quadratic numbers can be characterised as the complex numbers z for which the functions z and $j(z)$ simultaneously take algebraic values. The latter result has been greatly generalised by Shiga and Wolfart [223] and by Cohen [73]; they considered the map τ which associates with an abelian variety A of dimension g the normalised period matrix $\tau(A) = \Omega_1(A)\Omega_2(A)^{-1}$ in the Siegel upper half-plane $\mathfrak{H}_g$. Here $\Omega(A) = (\Omega_1(A), \Omega_2(A))$ is the matrix of period vectors of the abelian variety with respect to a basis for the space of translation invariant holomorphic 1-forms. They proved that the fields of definition of A and $\tau(A)$ are both algebraic if and only if A has complex multiplication by a CM-field; in this case $\tau(A)$ is diagonalisable and the diagonal elements together with their complex conjugates give a basis of a number field of degree $2g$ which is the associated CM-field.

If one considers the period map $\Omega(A)$ in non-normalised form rather than the normalised period matrix $\tau(A)$ as above, then one is led to the theory of hypergeometric functions and the problem of determining when the functions assume algebraic values is more difficult. Hypergeometric theory dates back to Gauss and it was extensively developed by Schwarz, Klein and Poincaré in the nineteenth century; their researches gave rise, in particular, to some important new geometric aspects. Recently the subject has been significantly applied by Deligne, Mostow and Margulis in connection with the arithmeticity of lattices, by Hirzebruch in studies relating to algebraic surfaces and by Gelfand, Kapranov and Zelevinsky in mathematical physics. In 1988 Wolfart [257] made the surprising discovery that certain classical hypergeometric functions have transcendence properties similar to those of the functions $e^{i\pi z}$ and $j(z)$ referred to above; we shall now give a brief description of this work, showing how it relates to the André–Oort conjecture and to the analytic subgroup theorem.

The hypergeometric function of Gauss is defined by

$$F(a,b,c;z) = 1 + \frac{ab}{c}z + \frac{a(a+1)b(b+1)}{2c(c+1)}z^2 + \cdots,$$

where a, b, c are real numbers with c not zero or a negative integer. In the case when a, b, c are rational, which we shall subsequently assume, there is a classical integral representation

$$F(a,b,c;z)\int_0^1 \omega(0) = \int_0^1 \omega(z),$$

where $\omega(z)$ is the rational differential form defined by

$$u^{b-1}(1-u)^{c-b-1}(1-zu)^{-a}du$$

and $\omega(0)$ is this differential form with $z = 0$. The function $F(a, b, c; z)$ is multivalued and satisfies the linear differential equation

$$z(1-z)\frac{d^2F}{dz^2} + (c - (a+b+1)z)\frac{dF}{dz} - abF = 0\,.$$

This is of Fuchsian type with singularities only at 0, 1 and ∞; we write for brevity $X = \mathbb{P}^1\backslash\{0, 1, \infty\}$ and as usual we denote by $X(\mathbb{C})$ the set of complex points of X. The solution space of the differential equation in a neighbourhood of each $z \in X(\mathbb{C})$ has dimension 2 and analytic continuation of a pair of linearly independent solutions along any closed path gives a new pair of solutions which can be expressed in terms of the original solution. This defines an invertible 2×2 matrix with algebraic integer coefficients and furnishes a representation of the fundamental group of $X = X(\mathbb{C})$ in $\mathrm{SL}_2(\mathbb{R})$; the representation is referred to as the monodromy representation and its image is called the monodromy group.

The differential form $\omega(z)$ determines a non-singular algebraic curve $C(z) = C(a, b, c; z)$ on which $\omega(z)$ is holomorphic, assuming some natural restrictions on a, b, c. If the genus of $C(z)$ is positive then the Jacobian contains a uniquely determined abelian subvariety $A(z) = A(a, b, c; z)$ with dimension $\varphi(q)$ where q is the least common denominator of a, b, c and where φ is the Euler totient-function. Similarly the differential form $\omega(0)$ defines an abelian variety $A(0) = A(a, b, c; 0)$ with dimension $\frac{1}{2}\varphi(q)$. Now, both the abelian varieties $A(z)$ and $A(0)$ have an endomorphism algebra which contains an order in the cyclotomic field $\mathbb{Q}(\zeta_q)$ with ζ_q a primitive qth root of unity. Further, the integral equation above can be expressed as a vanishing abelian logarithmic form and if both z and $F(a, b, c; z)$ are algebraic numbers then the associated abelian variety and the coefficients of the form are algebraic. The analytic subgroup theorem (see Theorem 6.1) now shows that there is a non-zero homomorphism from $A(z)$ onto $A(0)$ and this implies that $A(z)$ is isogenous to a product $A(0) \times B(0)$ where $B(0)$ is an abelian variety which, like $A(0)$, has complex multiplication by an order in $\mathbb{Q}(\zeta_q)$. Thus we deduce that the normalized period matrix $\tau(z) = \tau(A(z))$ in $\mathfrak{H}_g$ defines a special point in the Siegel modular variety associated with $\mathfrak{H}_g$ and the latter is contained in the image of X under the map τ. Since $\tau(X)$ is a curve, the Edixhoven–Yafaev theorem (see [74, 83]) implies that, if there are infinitely many z with both z and $F(a, b, c; z)$ algebraic, then $\tau(X)$ is a

Shimura curve. In this case the monodromy representation is arithmetic and, by a well-known theorem of Takeuchi, there are then precisely 85 triples (a, b, c); the corresponding hypergeometric functions are transcendental functions and, by Wolfart [257] and the remarks above, the z in question form a cyclotomic field. It remains to consider the case when the genus of $C(z)$ is 0; in this instance the differential form $\omega(z)$ is exact, the hypergeometric function is an algebraic function and, by a classical result of Schwarz, this is equivalent to the property that the monodromy group is finite. In summary, we see that the set of complex numbers z for which z and $F(a, b, c; z)$ are both algebraic is $\overline{\mathbb{Q}}$ if and only if the monodromy group is finite, it is a cyclotomic field if and only if the monodromy group is arithmetic and it is a finite set in all other cases.

8.6 The Manin–Mumford conjecture

In the 1960s, prior to Faltings' proof of the Mordell conjecture as discussed in Section 7.6, there were significant papers in this context by Manin [160] and by Mumford [183]. They were led to a question on torsion points on curves, regarded as embedded in their Jacobians, and this has come to be recognised as a research topic in Diophantine geometry in its own right. Specifically, the Manin–Mumford conjecture asserts that any irreducible curve defined over a field of characteristic 0 which contains infinitely many torsion points must necessarily be elliptic. The subject was taken up by Lang [136] and here he formulated a more general assertion which includes both the Manin–Mumford and the Mordell conjectures as special cases; in a refined form this states that if Γ is a subgroup of finite rank of a semi-abelian variety defined over an algebraically closed field L of characteristic 0 then the intersection of Γ with any closed subvariety V is contained in a finite union of translates of subgroup varieties all of which are contained in V. To deduce Mordell's conjecture from Lang's conjecture, we have to consider the case when L is an algebraic closure of a number field $\mathbb{K}$ and Γ is the group of $\mathbb{K}$-rational points of the Jacobian of a curve X of genus greater than 0. It follows from the Mordell–Weil theorem that Γ has finite rank and Lang's conjecture implies that the intersection $\Gamma \cap X(\mathbb{K})$ is contained in a finite union of abelian subvarieties in X; if this is infinite then the abelian subvarieties have dimension 1 whence the genus of X must be 1. Similarly one sees that Lang's conjecture implies the results of Vojta on subvarieties of semi-abelian varieties as referred to in Section 8.3. To deduce the Manin–Mumford conjecture from Lang's conjecture, we have to consider the case when Γ has rank 0; in this instance, Γ consists entirely of torsion points and one sees as above that any curve X with positive genus, regarded as embedded in its Jacobian, such that $\Gamma \cap X(L)$ is infinite must be elliptic.

Two distinct proofs of the Lang conjecture in the case when V is a curve and the semi-abelian variety is a torus, both of these assumed defined over a number field $\mathbb{K}$, were given by Lang and he attributes the underlying ideas to Ihara, Serre and Tate. Subsequently Laurent [142] established the conjecture subject only to the hypothesis that the semi-abelian variety is a torus and Bogomolov [44], returning to the number field case, studied the situation when the semi-abelian variety is assumed to be abelian. He succeeded in proving the assertion subject to restrictions on the order of the elements of Γ and his exposition involved studies relating to Galois representations on the Tate module; the latter provided a new tool in Diophantine geometry for transferring a geometric action into a Galois action and there are strong affinities here with the work of Serre on elliptic curves as mentioned in Section 7.5. The full Lang conjecture in the abelian case was proved by Raynaud [199, 200] and the general case was established in 1989 by Hindry [128]. Raynaud's method involved first reducing to the number field case and then studying the underlying abelian variety modulo p and p^2 for two distinct primes p of good reduction; this yields control of the full torsion group. Hindry, on the other hand, used a result of Serre on the image of the Galois group in the ring of endomorphisms of an abelian variety together with a technique from transcendence theory; the former had its origins in the paper of Bogomolov referred to earlier and the latter utilised a zero estimate originating from work of Masser and Wüstholz [165] in the context of fields with large transcendence degree.

References

[1] F. G. Acuña, H. Short, Cyclic branched coverings of knots and homology spheres, *Revista Mat. Madrid* **4** (1991), pp. 97–120.

[2] M. K. Agrawal, J. H. Coates, D. C. Hunt, A. J. van der Poorten, Elliptic curves of conductor 11, *Math. Comput.* **35** (1980), pp. 991–1002.

[3] F. Amoroso, U. Zannier (eds.), *Diophantine Approximation*, Lecture Notes in Mathematics **1819**, Berlin: Springer (2003).

[4] M. Anderson, Inhomogeneous linear forms in algebraic points of an elliptic function. In: *Transcendence Theory: Advances and Applications*, ed. A. Baker and D. W. Masser, London: Academic Press (1977), pp. 121–143.

[5] G. W. Anderson, W. D. Brownawell, M. A. Papanikolas, Determination of the algebraic relations among special Γ-values in positive characteristic, *Ann. Math.* **160** (2004), pp. 237–313.

[6] Y. André, Finitude des couples d'invariants modulaires singuliers sur une courbe algébrique plane non modulaire, *J. Reine Angew. Math.* **505** (1998), pp. 203–208.

[7] S. Arno, M. L. Robinson, F. S. Wheeler, Imaginary quadratic fields with small odd class number, *Acta Arithmetica* **83** (1998), pp. 295–330.

[8] V. I. Arnol'd, *Huygens and Barrow, Newton and Hook*, Basel: Birkhäuser (1990).

[9] M. F. Atiyah, I. G. Macdonald, *Introduction to Commutative Algebra*, Reading, MA: Addison-Wesley (1969).

[10] J. Ax, On the units of an algebraic number field, *Illinois J. Math.* **9** (1965), pp. 584–589.

[11] A. Baker, Continued fractions of transcendental numbers, *Mathematika* **9** (1962), pp. 1–8.

[12] A. Baker, Rational approximations to certain algebraic numbers, *Proc. London Math. Soc.* **14** (1964), pp. 385–398.

[13] A. Baker, Rational approximations to $\sqrt[3]{2}$ and other algebraic numbers, *Q. J. Math. Oxford* **15** (1964), pp. 375–383.
[14] A. Baker, On some Diophantine inequalities involving the exponential function, *Can. J. Math.* **17** (1965), pp. 616–626.
[15] A. Baker, Linear forms in the logarithms of algebraic numbers I, II, III, IV, *Mathematika* **13** (1966), pp. 204–216; **14** (1967), pp. 102–107; **14** (1967), pp. 220–228; **15** (1968), pp. 204–216.
[16] A. Baker, A note on integral integer-valued functions of several variables, *Proc. Cambridge Philos. Soc.* **63** (1967), pp. 715–720.
[17] A. Baker, Contributions to the theory of Diophantine equations I: On the representation of integers by binary forms; II: The Diophantine equation $y^2 = x^3 + k$, *Philos. Trans. R. Soc. London, Ser. A* **263** (1968), pp. 173–208.
[18] A. Baker, On the quasi-periods of the Weierstrass ζ-function, *Göttinger Nachr.* **16** (1969), pp. 145–157.
[19] A. Baker, Bounds for the solutions of the hyperelliptic equation, *Proc. Cambridge Philos. Soc.* **65** (1969), pp. 439–444.
[20] A. Baker, On the periods of the Weierstrass $\wp$-function, *Symposia Math. IV, INDAM, Rome, 1968*, London: Academic Press (1970), pp. 155–174.
[21] A. Baker, Imaginary quadratic fields with class number 2, *Ann. Math.* **94** (1971), pp. 139–152.
[22] A. Baker, On the class number of imaginary quadratic fields, *Bull. Am. Math. Soc.* **77** (1971), pp. 678–684.
[23] A. Baker, A sharpening for the bounds for linear forms in logarithms I, II, III, *Acta Arithmetica* **21** (1972), pp. 117–129; **24** (1973), pp. 33–36; **27** (1975), pp. 247–252.
[24] A. Baker, Some aspects of transcendence theory, *Astérisque* **24–25** (1975), pp. 169–175.
[25] A. Baker, *Transcendental Number Theory*, 1st edn, Cambridge: Cambridge University Press (1975), 3rd edn, Mathematical Library Series (1990).
[26] A. Baker, The theory of linear forms in logarithms. In: *Transcendence Theory: Advances and Applications*, ed. A. Baker and D. W. Masser, London: Academic Press (1977), pp. 1–27.
[27] A. Baker, Logarithmic forms and the *abc*-conjecture, *Number Theory (Eger, 1996)*, Berlin: de Gruyter (1998), pp. 37–44.
[28] A. Baker, Experiments on the *abc*-conjecture, *Publ. Math. Debrecen* **65** (2004), pp. 253–260.
[29] A. Baker, J. Coates, Integer points on curves of genus 1, *Proc. Cambridge Philos. Soc.* **67** (1970), pp. 595–602.
[30] A. Baker, H. Davenport, The equations $3x^2 - 2 = y^2$ and $8x^2 - 7 = z^2$, *Q. J. Math. Oxford* **20** (1969), pp. 129–137.

[31] A. Baker, H. M. Stark, On a fundamental inequality in number theory, *Ann. Math.* **94** (1971), pp. 190–199.

[32] A. Baker, C. L. Stewart, On effective approximations to cubic irrationals. In: *New Advances in Transcendence Theory*, ed. A. Baker, Cambridge: Cambridge University Press (1988), pp. 1–24.

[33] A. Baker, G. Wüstholz, Logarithmic forms and group varieties, *J. Reine Angew. Math.* **442** (1993), pp. 19–62.

[34] A. Baker, G. Wüstholz, Number theory, transcendence and Diophantine geometry in the next millennium, *Mathematics: Frontiers and Perspectives*, Providence, RI: American Mathematical Society (2000), pp. 1–12.

[35] A. Baker, B. J. Birch, E. Wirsing, On a problem of Chowla, *J. Number Theory* **5** (1973), pp. 224–236.

[36] K. Barré-Sirieix, G. Diaz, F. Gramain, G. Philibert, Une preuve de la conjecture de Mahler–Manin, *Invent. Math.* **124** (1996), pp. 1–9.

[37] G.V. Belyĭ, On the Galois extensions of a maximal cyclotomic field, *Izv. Akad. Nauk SSSR Ser. Mat.* **43** (1979), pp. 267–276.

[38] D. Bertrand, Galois representations and transcendental numbers. In: *New Advances in Transcendence Theory*, ed. A. Baker, Cambridge: Cambridge University Press (1988), pp. 25–36.

[39] D. Bertrand, Theta functions and transcendence, *Ramanujan J.* **1** (1997), pp. 339–350.

[40] D. Bertrand, D. W. Masser, Linear forms in elliptic integrals, *Invent. Math.* **58** (1980), pp. 283–288.

[41] Yu. Bilu, Baker's method and modular curves. In: *A Panorama of Number Theory or the View from Baker's Garden*, ed. G. Wüstholz, Cambridge: Cambridge University Press (2002), pp. 73–88, .

[42] Yu. Bilu, Catalan's conjecture (after Mihăilescu), Séminaire Bourbaki **909** (2002–2003), *Astérisque* **294** (2004), pp. 1–26.

[43] Yu. Bilu, G. Hanrot, P. M. Voutier, Existence of primitive divisors of Lucas and Lehmer numbers, *J. Reine Angew. Math.* **539** (2001), pp. 75–122.

[44] F. A. Bogomolov, Points of finite order on abelian varieties, *Izv. Akad. Nauk SSSR Ser. Mat.* **44** (1980), pp. 782–804, 973.

[45] E. Bombieri, Algebraic values of meromorphic maps, *Invent. Math.* **10** (1970), pp. 267–287.

[46] E. Bombieri, Counting points on curves over finite fields (d'après S. A. Stepanov), *Séminaire Bourbaki* **430** (1972/1973), Lecture Notes in Mathematics **383**, Berlin: Springer (1974), pp. 234–241.

[47] E. Bombieri, Effective Diophantine approximation on G_m, *Ann. Scuola Norm. Sup. Pisa Cl. Sci.* **20** (1993), pp. 61–89.

[48] E. Bombieri, W. Gubler, *Heights in Diophantine Geometry*, New Mathematical Monographs **4**, Cambridge: Cambridge University Press (2006).

[49] E. Bombieri, S. Lang, Analytic subgroups of group varieties, *Invent. Math.* **11** (1970), pp. 1–14.

[50] E. Bombieri, J. Vaaler, On Siegel's lemma, *Invent. Math.* **73** (1983), pp. 11–32; Addendum, **75** (1984), p. 377.

[51] A. Borel, *Linear Algebraic Groups*, New York: W. A. Benjamin (1969).

[52] W. Bosma, H. W. Lenstra, Complete systems of two addition laws for elliptic curves, *J. Number Theory* **53** (1995), pp. 229–240.

[53] V. Bosser, Minorations de formes linéaires de logarithmes pour les modules de Drinfeld, *J. Number Theory* **75** (1999), pp. 279–323.

[54] J.-B. Bost, Périodes et isogénies des variétés abéliennes sur les corps de nombres, *Astérisque* **237** (1996), pp. 115–161.

[55] B. Boyer, *A History of Mathematics*, New York: Wiley (1968).

[56] B. Brindza, On S-integral solutions of the equation $y^m = f(x)$, *Acta Math. Hung.* **44** (1984), pp. 133–139.

[57] B. Brindza, On some generalizations of the Diophantine equation $1^k + 2^k + \cdots + x^k = y^z$, *Acta Arithmetica* **44** (1984), pp. 99–107.

[58] B. Brindza, On the equation $f(x) = y^m$ over finitely generated domains, *Acta Math. Hung.* **53** (1989), pp. 377–383.

[59] E. Brown, Sets in which $xy + k$ are always squares, *Math. Comput.* **45** (1985), pp. 613–620.

[60] W. D. Brownawell, Minimal extensions of algebraic groups and linear independence, *J. Number Theory* **90** (2001), pp. 239–254.

[61] W. D. Brownawell, D. W. Masser, Multiplicity estimates for analytic functions I, II, *J. Reine Angew. Math.* **314** (1980), pp. 200–216; *Duke Math. J.* **47** (1980), pp. 273–295.

[62] A. Brumer, On the units of algebraic number fields, *Mathematika* **14** (1967), pp. 121–124.

[63] J. W. S. Cassels, *An Introduction to Diophantine Approximation*, Cambridge: Cambridge University Press (1957).

[64] J. W. S. Cassels, *An Introduction to the Geometry of Numbers*, 2nd edn, Grundlehren Math. **99**, Berlin: Springer (1971).

[65] D. V. Chudnovsky, G. V. Chudnovsky, Padé approximations and Diophantine geometry, *Proc. Natl. Acad. Sci. USA* **82** (1985), pp. 2212–2216.

[66] D. V. Chudnovsky, G. V. Chudnovsky, Approximations and complex multiplication according to Ramanujan, *Ramanujan revisited (Urbana-Champaign, Ill., 1987)*, Boston, MA: Academic Press (1988), pp. 375–472.

[67] G. V. Chudnovsky, Algebraic independence of several values of the exponential function, *Mat. Zametki* **15** (1974), pp. 661–672.

[68] G. V. Chudnovsky, Algebraic independence of the values of elliptic functions at algebraic points, *Invent. Math.* **61** (1980), pp. 267–290.

[69] G. V. Chudnovsky, On the method of Thue–Siegel, *Ann. Math.* **117** (1983), pp. 325–382.

[70] J. Coates, An effective p-adic analogue of a theorem of Thue I, *Acta Arithmetica* **15** (1969), pp. 279–305.

[71] J. Coates, The transcendence of linear forms in $\omega_1, \omega_2, \eta_1, \eta_2, 2\pi i$, *Am. J. Math.* **93** (1971), pp. 385–397.

[72] P. Cohen, On the coefficients of the transformation polynomials for the elliptic modular function, *Math. Proc. Cambridge Philos. Soc.* **95** (1984), pp. 389–402.

[73] P. Cohen, Humbert surfaces and transcendence properties of automorphic functions, *Rocky Mountain J. Math.* **26** (1996), pp. 987–1001.

[74] P. Cohen, G. Wüstholz, Application of the André–Oort conjecture to some questions in transcendence. In: *A Panorama of Number Theory or the View from Baker's Garden*, ed. G. Wüstholz, Cambridge: Cambridge University Press (2002), pp. 89–106.

[75] S. David, Minorations de formes linéaires de logarithmes elliptiques, *Mém. Soc. Math. Fr.* **62** (1995).

[76] S. David, N. Hirata-Kohno, Recent progress on linear forms in elliptic logarithms. In: *A Panorama of Number Theory or the View from Baker's Garden*, ed. G. Wüstholz, Cambridge: Cambridge University Press (2002), pp. 26–37.

[77] C. S. Davis, Rational approximations to e, *J. Aust. Math. Soc. Ser. A* **25** (1978), pp. 497–502.

[78] E. Dobrowolski, On a question of Lehmer and the number of irreducible factors of a poynomial, *Acta Arithmetica* **34** (1979), pp. 391–401.

[79] A. Dujella, A proof of the Hoggatt–Bergum conjecture, *Proc. Am. Math. Soc.* **127** (1999), pp. 1999–2005.

[80] A. Dujella, There are only finitely many Diophantine quintuples, *J. Reine Angew. Math.* **566** (2004), pp. 183–214.

[81] F. J. Dyson, The approximation to algebraic numbers by rationals, *Acta Math.* **79** (1947), pp. 225–240.

[82] B. Edixhoven, On the André–Oort conjecture for Hilbert modular surfaces, Moduli of abelian varieties (Texel Island 1999), *Prog. Math.* **195** (2001), pp. 111–155.

[83] B. Edixhoven, A. Yafaev, Subvarieties of Shimura varieties, *Ann. Math.* **157** (2003), pp. 621–645.

[84] W. J. Ellison, F. Ellison, J. Pesek, C. E. Stahl, D. S. Stall, The Diophantine equation $y^2 + k = x^3$, *J. Number Theory* **4** (1972), pp. 107–117.

[85] P. Erdős, On the irrationality of certain series: problems and results. In: *New Advances in Transcendence Theory*, ed. A. Baker, Cambridge: Cambridge University Press (1988), pp. 102–109.

[86] J.-H. Evertse, Points on subvarieties of tori. In: *A Panorama of Number Theory or the View from Baker's Garden*, ed. G. Wüstholz, Cambridge: Cambridge University Press (2002), pp. 214–230.

[87] J.-H. Evertse, H. P. Schlickewei, A quantitative version of the absolute subspace theorem, *J. Reine Angew. Math.* **548** (2002), pp. 21–127.

[88] J.-H. Evertse, K. Győry, C. L. Stewart, R. Tijdeman, S-unit equations and their applications. In: *New Advances in Transcendence Theory*, ed. A. Baker, Cambridge: Cambridge University Press (1988), pp. 110–174.

[89] J.-H. Evertse, H. P. Schlickewei, W. M. Schmidt, Linear equations in variables which lie in a multiplicative group, *Ann. Math.* **155** (2002), pp. 807–836.

[90] G. Faltings, Endlichkeitssätze für abelsche Varietäten über Zahlkörpern, *Invent. Math.* **73** (1983), pp. 349–366.

[91] G. Faltings, Diophantine approximation on abelian varieties, *Ann. Math.* **133** (1991), pp. 549–576.

[92] G. Faltings, G. Wüstholz, Einbettungen kommutativer algebraischer Gruppen und einige ihrer Eigenschaften, *J. Reine Angew. Math.* **354** (1984), pp. 176–205.

[93] G. Faltings, G. Wüstholz, Diophantine approximations on projective spaces, *Invent. Math.* **116** (1994), pp. 109–138.

[94] G. Faltings, G. Wüstholz, F. Grunewald, N. Schappacher, U. Stuhler, *Rational Points*, 3rd edn, Aspects of Mathematics, E6, Berlin: Vieweg (1992).

[95] N. I. Feldman, Estimation of a linear form in the logarithms of algebraic numbers, *Mat. Sb.* **76** (1968), pp. 304–319; and similar titles, *Mat. Sb.* **77** (1968), pp. 423–436; *Usp. Mat. Nauk* **23** (1968), pp. 185–186; *Dokl. Akad. Nauk* **182** (1968), pp. 1278–1279.

[96] N. I. Feldman, An inequality for a linear form in logarithms of algebraic numbers, *Mat. Zametki* **5** (1969), pp. 681–689.

[97] N. I. Feldman, Refinement of two effective inequalities of A. Baker, *Mat. Zametki* **6** (1969), pp. 767–769.

[98] N. I. Feldman, An effective refinement of the exponent in Liouville's theorem, *Izv. Akad. Nauk* **35** (1971), pp. 973–990.

[99] R. Ferretti, An effective version of Faltings product theorem, *Forum Math.* **8** (1996), pp. 401–427.

[100] S. Fukasawa, Über ganzwertige ganze Funktionen, *Tôhoku Math. J.* **27** (1926), pp. 41–52.

[101] É. Gaudron, Mesure d'indépendance linéaire de logarithmes dans un groupe algébrique commutatif, *C. R. Math. Acad. Sci. Paris* **333** (2001), pp. 1059–1064; *Invent. Math.* **162** (2005), pp. 137–188.

[102] J. Gebel, A. Pethő, H. G. Zimmer, Computing integer points on elliptic curves, *Acta Arithmetica* **68** (1994), pp. 171–192.

[103] J. Gebel, A. Pethő, H. G. Zimmer, On Mordell's equation, *Compos. Math.* **110** (1998), pp. 335–367.

[104] A. O. Gelfond, Sur les propriétés arithmétiques des fonctions entières, *Tôhoku Math. J.* **30** (1929), pp. 280–285.

[105] A. O. Gelfond, Sur le septième problème de Hilbert, *Izv. Akad. Nauk SSSR* **7** (1934), pp. 623–630; *Dokl. Akad. Nauk* **2** (1934), pp. 1–6.

[106] A. O. Gelfond, On the approximation of transcendental numbers by algebraic numbers, *Dokl. Akad. Nauk* **2** (1935), pp. 177–182.

[107] A. O. Gelfond, On the approximation by algebraic numbers to the ratio of the logarithms of two algebraic numbers, *Izv. Akad. Nauk SSSR* **5–6** (1939), pp. 509–518.

[108] A. O. Gelfond, The approximation of algebraic numbers by algebraic numbers and the theory of transcendental numbers, *Usp. Mat. Nauk* **4** (1949), pp. 19–49.

[109] A. O. Gelfond, On the algebraic independence of algebraic powers of algebraic numbers, *Dokl. Akad. Nauk* **64** (1949), pp. 277–280; and similar titles, *Dokl. Akad. Nauk* **67** (1949), pp. 13–14; *Usp. Mat. Nauk* **5** (1949), pp. 40–48; see also *Am. Math. Soc. Transl.* **2** (1962), pp. 125–169.

[110] A. O. Gelfond, *Transcendental and Algebraic Numbers*, New York: Dover (1960).

[111] A. O. Gelfond, Yu. V. Linnik, On Thue's method and the problems of effectiveness in quadratic fields, *Dokl. Akad. Nauk* **61** (1948), pp. 773–776.

[112] A. O. Gelfond, Yu. V. Linnik, *Elementary Methods in Analytic Number Theory*, Chicago, IL: Rand McNally (1965).

[113] C. I. Gerhardt (ed.), *Leibnizens Mathematische Schriften*, Erste Abtheilung, Bd. II, Berlin: Ascher (1850).

[114] D. M. Goldfeld, The class number of quadratic fields and the conjectures of Birch and Swinnerton-Dyer, *Ann. Scuola Norm. Sup. Pisa Cl. Sci.* (4), **3** (1976), pp. 624–663; see also *Astérisque* **41–42** (1977).

[115] D. M. Goldfeld, Gauss' class number problem for imaginary quadratic fields, *Bull. Am. Math. Soc.* **13** (1985), pp. 23–37.

[116] F. Gramain, Sur le théorème de Fukasawa–Gelfond–Gruman–Masser, *Séminaire Delange–Pisot–Poitou 1980–1981*, Basel: Birkhäuser (1981).

[117] A. Granville, *ABC* allows us to count squarefrees, *Int. Math. Res. Not.* **19** (1998), pp. 991–1009.

[118] A. Granville, H. M. Stark, *abc* implies no 'Siegel zeros' for *L*-functions of characters with negative discriminant, *Invent. Math.* **139** (2000), pp. 509–523.

[119] A. Granville, T. J. Tucker, It's as easy as *abc*, *Not. Am. Math. Soc.* **49** (2002), pp. 1224–1231.

[120] G. M. Grinstead, On a method of solving a class of Diophantine equations, *Math. Comput.* **32** (1978), pp. 936–940.

[121] B. Gross, D. Zagier, Heegner points and derivatives of L-series, *Invent. Math.* **84** (1986), pp. 225–320.

[122] R. Güting, Approximation of algebraic numbers by algebraic numbers, *Michigan Math. J.* **8** (1961), pp. 149–159.

[123] K. Győry, Solving Diophantine equations by Baker's theory. In: *A Panorama of Number Theory or the View from Baker's Garden*, ed. G. Wüstholz, Cambridge: Cambridge University Press (2002), pp. 38–72.

[124] K. Győry, R. Tijdeman, M. Voorhoeve, On the equation $1^k + 2^k + \cdots + x^k + R(x) = y^z$, *Acta Math.* **143** (1979), pp. 1–8.

[125] R. Hartshorne, *Algebraic Geometry*, Graduate Texts in Mathematics **52**, Berlin: Springer (1977).

[126] K. Heegner, Diophantische Analysis und Modulfunktionen, *Math. Z.* **56** (1952), pp. 227–253.

[127] C. Hermite, Sur la fonction exponentielle, *Comptes Rendus* **77** (1873), pp. 18–24, 74–79, 226–233; see also *Oeuvres* III, pp. 150–181.

[128] M. Hindry, Autour d'une conjecture de Serge Lang, *Invent. Math.* **94** (1988), pp. 575–603.

[129] N. Hirata-Kohno, Formes linéaires de logarithmes de points algébriques sur les groupes algébriques, *Invent. Math.* **104** (1991), pp. 401–433.

[130] D. Y. Kleinbock, G. A. Margulis, Flows on homogeneous spaces and Diophantine approximation on manifolds, *Ann. Math.* **148** (1998), pp. 339–360.

[131] J. Kollar, Effective Nullstellensatz for arbitrary ideals, *J. Eur. Math. Soc.* **1** (1999), pp. 313–337.

[132] R. O. Kuzmin, On a new class of transcendental numbers, *Izv. Akad. Nauk SSSR* **3** (1930), pp. 583–597.

[133] J. H. Lambert, Mémoire sur quelques propriétés remarquables des quantités transcendantes circulaires et logarithmiques, *Histoire Acad. R. Sci. Belles Lettres, Berlin*, Année 1761 (1768), pp. 265–322; see also *Opera Math.* II, pp. 112–159.

[134] S. Lang, Transcendental points on group varieties, *Topology* **1** (1962), pp. 313–318.

[135] S. Lang, Diophantine approximations on toruses, *Am. J. Math.* **86** (1964), pp. 521–533.

[136] S. Lang, Division points on curves, *Ann. Mat. Pura Appl.* **70** (1965), pp. 229–234.

[137] S. Lang, *Introduction to Transcendental Numbers*, Reading, MA: Addison Wesley (1966).

[138] S. Lang, *Number Theory III: Diophantine Geometry*, Berlin: Springer (1991).

[139] S. Lang, Die *abc*-Vermutung, *Elem. Math.* **48** (1993), pp. 89–99.

[140] H. Lange, W. Ruppert, Complete systems of addition laws on abelian varieties, *Invent. Math.* **79** (1985), pp. 603–610.

[141] M. Laurent, Transcendance de périodes d'intégrales elliptiques I, II, *J. Reine Angew. Math.* **316** (1980), pp. 123–139; **333** (1982), pp. 144–161.

[142] M. Laurent, Equations diophantiennes exponentielles, *Invent. Math.* **78** (1984), pp. 299–327.

[143] M. Laurent, Linear forms in two logarithms and interpolation determinants, *Acta Arithmetica* **66** (1994), pp. 181–199.

[144] A. K. Lenstra, H. W. Lenstra, L. Lovász, Factoring polynomials with rational coefficients, *Math. Ann.* **261** (1982), pp. 515–534.

[145] W. J. LeVeque, *Topics in Number Theory II*, 1st edn, Reading, MA: Addison-Wesley (1956), 2nd edn, Mineola, NY: Dover Publications (2002).

[146] F. Lindemann, Über die Zahl π, *Math. Ann.* **20** (1882), pp. 213–225.

[147] J. Liouville, Sur des classes très-étendues de quantités dont la valeur n'est ni algébrique, ni même reductible à des irrationelles algébriques, *Comptes Rendus*. **18** (1844), pp. 883–885, 910–911; *J. Math. Pures Appl.* **16** (1851), pp. 133–142.

[148] T. Loher, D. Masser, Uniformly counting points of bounded height, *Acta Arithmetica* **111** (2004), pp. 277–297.

[149] J. H. Loxton, Automata and transcendence. In: *New Advances in Transcendence Theory*, ed. A. Baker, Cambridge: Cambridge University Press (1988), pp. 215–228.

[150] J. H. Loxton, A. J. van der Poorten, Transcendence and algebraic independence by a method of Mahler. In: *Transcendence Theory: Advances and Applications*, ed. A. Baker and D. W. Masser, London: Academic Press (1977), pp. 211–226.

[151] K. Mahler, Arithmetische Eigenschaften der Lösungen einer Klasse von Funktionalgleichungen, *Math. Ann.* **101** (1929), pp. 342–366.

[152] K. Mahler, Über das Verschwinden von Potenzreihen mehrerer Veränderlicher in speziellen Punktfolgen, *Math. Ann.* **103** (1930), pp. 573–587.

[153] K. Mahler, Arithmetische Eigenschaften einer Klasse von transzendental-transzendenter Funktionen, *Math. Z.* **32** (1930), pp. 545–585.

[154] K. Mahler, Über transzendente *p*-adische Zahlen, *Compos. Math.* **2** (1935), pp. 259–275.

[155] K. Mahler, On the approximation of π, *Indagat. Math.* **15** (1953), pp. 30–42.

[156] K. Mahler, Remarks on a paper by W. Schwarz, *J. Number Theory* **1** (1969), pp. 512–521.

[157] K. Mahler, On the coefficients of transformation polynomials for the modular function, *Bull. Aust. Math. Soc.* **10** (1974), pp. 197–218.

[158] K. Mahler, *Lectures on Transcendental Numbers*, Lecture Notes in Mathematics **546**, Berlin: Springer (1976).

[159] E. Maillet, *Introduction à la Théorie des Nombres Transcendants et de Propriétés Arithmétiques des Fonctions*, Paris: Gauthier-Villars (1906).

[160] Yu. Manin, Rational points on algebraic curves over function fields, *Izv. Akad. Nauk. SSSR Ser. Mat.* **27** (1963), pp. 1395–1440.

[161] R. C. Mason, *Diophantine Equations over Function Fields*, London Math. Soc. Lecture Notes **96**, Cambridge: Cambridge University Press (1984).

[162] D. W. Masser, *Elliptic Functions and Transcendence*, Lecture Notes in Mathematics **437**, Berlin: Springer (1975).

[163] D. W. Masser, Open problems. In: *Proc. Symp. Analytic Number Theory*, ed. W. W. L. Chen, London: Imperial College (1985).

[164] D. W. Masser, G. Wüstholz, Zero estimates on group varieties I, *Invent. Math.* **64** (1981), pp. 489–516.

[165] D. W. Masser, G. Wüstholz, Fields of large transcendence degree generated by values of elliptic functions, *Invent. Math.* **72** (1983), pp. 407–464.

[166] D. W. Masser, G. Wüstholz, Zero estimates on group varieties II, *Invent. Math.* **80** (1985), pp. 233–267.

[167] D. W. Masser, G. Wüstholz, Some effective estimates for elliptic curves, *Arithmetic of Complex Manifolds*, Lecture Notes in Mathematics **1399**, Berlin: Springer (1989), pp. 103–109.

[168] D. W. Masser, G. Wüstholz, Estimating isogenies on elliptic curves, *Invent. Math.* **100** (1990), pp. 1–24.

[169] D. W. Masser, G. Wüstholz, Galois properties of division fields of elliptic curves, *Bull. London Math. Soc.* **25** (1992), pp. 247–254.

[170] D. W. Masser, G. Wüstholz, Periods and minimal abelian subvarieties, *Ann. Math.* **137** (1993), pp. 407–458.

[171] D. W. Masser, G. Wüstholz, Isogeny estimates for abelian varieties, and finiteness theorems, *Ann. Math.* **137** (1993), pp. 459–472.

[172] D. W. Masser, G. Wüstholz, Endomorphism estimates for abelian varieties, *Math. Z.* **215** (1994), pp. 641–653.

[173] D. W. Masser, G. Wüstholz, Factorization estimates for abelian varieties, *Publ. Math. IHES* **81** (1995), pp. 5–24.

[174] E. M. Matveev, An explicit lower bound for a homogeneous rational linear form in logarithms of algebraic numbers I, II, *Izv. Math.* **62** (1998), pp. 723–772; **64** (2000), pp. 1217–1269.

[175] R. B. McFeat, Geometry of numbers in adele spaces, *Dissertationes Math. (Rozpr. Mat.)* **88** (1971).

[176] M. Meyer, A. Pajor, Sections of the unit ball of l_p^n, *J. Functional Analysis* **80** (1988), pp. 109–123.

[177] P. Mihăilescu, A class number free criterion for Catalan's conjecture, *J. Number Theory* **99** (2003), pp. 225–231.

[178] P. Mihăilescu, Primary cyclotomic units and a proof of Catalan's conjecture, *J. Reine Angew. Math.* **572** (2004), pp. 167–195.

[179] P. Mihăilescu, On the class groups of cyclotomic extensions in presence of a solution to Catalan's equation, *J. Number Theory* **118** (2006), pp. 123–144.

[180] M. Mkaouar, Continued fractions of transcendental numbers, *Bull. Greek Math. Soc.* **45** (2001), pp. 79–85.

[181] L. J. Mordell, On the rational solutions of the indeterminate equations of the third and fourth degrees, *Proc. Cambridge Philos. Soc.* **21** (1922), pp. 179–192.

[182] L. J. Mordell, *Diophantine Equations*, London: Academic Press (1969).

[183] D. Mumford, A remark on Mordell's conjecture, *Am. J. Math.* **87** (1965), pp. 1007–1016.

[184] D. Mumford, *Abelian Varieties*, 2nd edn, Tata Institute, Oxford: Oxford University Press (1974).

[185] M. R. Murty, The Ramanujan τ-function, *Ramanujan Revisited*, Academic Press (1988), pp. 269–288.

[186] Yu. V. Nesterenko, Bounds on the order of zeros of a class of functions and their application to the theory of transcendental numbers, *Math. USSR Izv.* **11** (1977), pp. 253–284.

[187] Yu. V. Nesterenko, P. Philippon (eds.), *Introduction to Algebraic Independence Theory*, Lecture Notes in Mathematics **1752**, Berlin: Springer (2001).

[188] K. Nishioka, *Mahler Functions and Transcendence*, Lecture Notes in Mathematics **1631**, Berlin: Springer (1996).

[189] R. W. K. Odoni, An application of the S-unit theorem to modular forms on $\Gamma_0(N)$. In: *New Advances in Transcendence Theory*, ed. A. Baker, Cambridge: Cambridge University Press (1988), pp. 270–279.

[190] F. Pellarin, Sur une majoration explicite pour un degré d'isogénie liant deux courbes elliptiques, *Acta Arithmetica* **100** (2001), pp. 203–243.

[191] A. Pethő, Full cubes in the Fibonacci sequence, *Publ. Math. Debrecen* **30** (1983), pp. 117–127.

[192] P. Philippon, Variétés abéliennes et indépendance algébrique. II. Un analogue abélien du théorème de Lindemann–Weierstrass, *Invent. Math.* **72** (1983), pp. 389–405.

[193] P. Philippon, Lemme de zéros dans les groupes algébriques commutatifs, *Bull. Soc. Math. Fr.* **114** (1986), pp. 355–383; Errata and addenda, **115** (1987), pp. 397–398.

[194] P. Philippon, M. Waldschmidt, Lower bounds for linear forms in logarithms. In: *New Advances in Transcendence Theory*, ed. A. Baker, Cambridge: Cambridge University Press (1988), pp. 280–312.

[195] R. G. E. Pinch, Simultaneous Pellian equations, *Math. Proc. Cambridge Philos. Soc.* **103** (1988), pp. 35–46.

[196] G. Pólya, Über ganze Funktionen, *Rend. Circ. Math. Palermo* **40** (1915), pp. 1–16.

[197] A. J. van der Poorten, On Baker's inequality for linear forms in logarithms, *Math. Proc. Cambridge Philos. Soc.* **80** (1976), pp. 233–248.

[198] A. J. van der Poorten, Linear forms in logarithms in the p-adic case. In: *Transcendence Theory: Advances and Applications*, ed. A. Baker and D. W. Masser, London: Academic Press (1977), pp. 29–57.

[199] M. Raynaud, Sous-variétés d'une variété abélienne et points de torsion, *Arithmetic and geometry I, Progr. Math.* **35**, Boston, MA: Birkhäuser (1983) pp. 327–352.

[200] M. Raynaud, Courbes sur une variété abélienne et points de torsion, *Invent. Math.* **71** (1983), pp. 207–233.

[201] P. Ribenboim, *Catalan's Conjecture*, Boston, MA: Academic Press (1994).

[202] R. Riley, Growth of order of homology of cyclic branched covers of knots, *Bull. London Math. Soc.* **22** (1990), pp. 287–297.

[203] J. B. Rosser, L. Schoenfeld, Approximate formulas for some functions of prime numbers, *Illinois J. Math.* **6** (1962), pp. 64–94.

[204] K. F. Roth, Rational approximations to algebraic numbers, *Mathematika* **2** (1955), pp. 1–20; Corrigendum, p. 168.

[205] J. J. Schäffer, The equation $1^p + 2^p + 3^p + \cdots + n^p = m^q$, *Acta Math.* **95** (1956) pp. 155–189.

[206] A. Schinzel, R. Tijdeman, On the equation $y^m = P(x)$, *Acta Arithmetica* **31** (1976), pp. 199–204.

[207] K. Schmidt, *Dynamical Systems of Algebraic Origin*, Basel: Birkhäuser (1995).

[208] W. M. Schmidt, *Equations Over Finite Fields: an Elementary Approach*, 1st edn, Lecture Notes in Mathematics **536**, Berlin: Springer (1976), 2nd edn, Heber City, UT: Kendrick Press (2004).

[209] W. M. Schmidt, *Diophantine Approximation*, Lecture Notes in Mathematics **785**, Berlin: Springer (1980).

[210] W. M. Schmidt, The subspace theorem in Diophantine approximations, *Compos. Math.* **69** (1989), pp. 121–173.

[211] T. Schneider, Transzendenzuntersuchungen periodischer Funktionen: I Transzendenz von Potenzen; II Transzendenzeigenschaften elliptischer Funktionen, *J. Reine Angew. Math.* **172** (1934), pp. 65–74.

[212] T. Schneider, Über die Approximation algebraischer Zahlen, *J. Reine Angew. Math.* **175** (1936), pp. 182–92.

[213] T. Schneider, Arithmetische Untersuchungen elliptischer Integrale, *Math. Ann.* **113** (1937), pp. 1–13.

[214] T. Schneider, Zur Theorie der Abelschen Funktionen und Integrale, *J. Reine Angew. Math.* **183** (1941), pp. 110–128.

[215] T. Schneider, Ein Satz über ganzwertige Funktionen als Prinzip für Transzendenzbeweise, *Math. Ann.* **121** (1949), pp. 131–140.

[216] T. Schneider, *Einführung in die transzendenten Zahlen*, Berlin: Springer (1957).

[217] J.-P. Serre, *Abelian ℓ-adic Representations and Elliptic Curves*, New York: Benjamin (1968).

[218] J.-P. Serre, Propriétés galoisiennes des points d'ordre fini des courbes elliptiques, *Invent. Math.* **15** (1972), pp. 259–331.

[219] J.-P. Serre, Quelques applications du théorème de densité de Chebotarev, *Publ. Math. IHES* **54** (1981), pp. 123–201.

[220] J.-P. Serre, *Algebraic Groups and Class Fields*, Berlin: Springer (1988).

[221] A. B. Shidlovsky, On a criterion for algebraic independence of the values of a class of integral functions, *Izv. Akad. Nauk SSSR* **23** (1959), pp. 35–66; see also *Am. Math. Soc. Transl.* **22** (1962), pp. 339–370.

[222] A. B. Shidlovsky, *Transcendental Numbers*, Studies in Mathematics **12**, Berlin: de Gruyter (1988).

[223] H. Shiga, J. Wolfart, Criteria for complex multiplication and transcendence properties of automorphic functions, *J. Reine Angew. Math.* **463** (1995), pp. 1–25.

[224] T. N. Shorey, On linear forms in logarithms of algebraic numbers, *Acta Arithmetica* **30** (1976), pp. 27–42.

[225] T. N. Shorey, R. Tijdeman, *Exponential Diophantine Equations*, Cambridge: Cambridge University Press (1986).

[226] C. L. Siegel, Approximation algebraischer Zahlen, *Math. Z.* **10** (1921), pp. 173–213.

[227] C. L. Siegel (under the pseudonym X), The integer solutions of the equation $y^2 = ax^n + bx^{n-1} + \cdots + k$, *J. London Math. Soc.* **1** (1926), pp. 66–68.

[228] C. L. Siegel, Über einige Anwendungen diophantischer Approximationen, *Abh. Preuss. Akad. Wiss.* **1** (1929).

[229] C. L. Siegel, Über die Perioden elliptischer Funktionen, *J. Reine Angew. Math.* **167** (1932), pp. 62–69.

[230] C. L. Siegel, Über die Classenzahl quadratischer Zahlkörper, *Acta Arithmetica* **1** (1936), pp. 83–86.

[231] C. L. Siegel, *Transcendental Numbers*, Princeton, NJ: Princeton University Press (1948).

[232] N. P. Smart, *The Algorithmic Resolution of Diophantine Equations*, Cambridge: Cambridge University Press (1998).

[233] V. G. Sprindžuk, Concerning Baker's theorem on linear forms in logarithms, *Dokl. Akad. Nauk* **11** (1967), pp. 767–769.

[234] V. G. Sprindžuk, *Classical Diophantine Equations*, Lecture Notes in Mathematics **1559**, Berlin: Springer (1982).

[235] H. M. Stark, A complete determination of the complex quadratic fields of classnumber one, *Michigan Math. J.* **14** (1967), pp. 1–27.

[236] H. M. Stark, A transcendence theorem for class number problems, *Ann. Math.* **94** (1971), pp. 153–173.

[237] C. L. Stewart, R. Tijdeman, On the Oesterlé–Masser conjecture, *Monatsh. Math.* **102** (1986), pp. 251–257.

[238] C. L. Stewart, Kunrui Yu, On the *abc*-conjecture I, II, *Math. Ann.* **291** (1991), pp. 225–230; *Duke Math. J.* **108** (2001), pp. 169–181.

[239] W. W. Stothers, Polynomial identities and Hauptmoduln, *Q. J. Math. Oxford* **32** (1981), pp. 349–370.

[240] R. J. Stroeker, N. Tzanakis, Solving elliptic Diophantine equations by estimating linear forms in elliptic logarithms, *Acta Arithmetica* **67** (1994), pp. 177–196.

[241] J. Tate, Endomorphisms of abelian varieties over finite fields, *Invent. Math.* **2** (1966), pp. 134–144.

[242] D. Thakur, *Function Field Arithmetic*, River Edge, NJ: World Scientific (2004).

[243] A. Thue, Über Annäherungswerte algebraischer Zahlen, *J. Reine Angew. Math.* **135** (1909), pp. 284–305.

[244] R. Tijdeman, On the equation of Catalan, *Acta Arithmetica* **29** (1976), pp. 197–209.

[245] R. Tijdeman, *Diophantine Equations and Diophantine Approximation*, Leiden: Math. Inst. Univ. Leiden (1988).

[246] N. Tzanakis, B. M. M. de Weger, On the practical solution of the Thue equation, *J. Number Theory* **31** (1989), pp. 99–132.

[247] J. D. Vaaler, A geometric inequality with applications to linear forms, *Pacific J. Math.* **83** (1979), pp. 543–553.

[248] A. I. Vinogradov, V. G. Sprindžuk, The representation of numbers by binary forms, *Mat. Zametki* **3** (1968), pp. 369–376.

[249] P. Vojta, *Diophantine Approximations and Value Distribution Theory*, Lecture Notes in Mathematics **1239**, Berlin: Springer (1987).

[250] P. Vojta, Siegel's theorem in the compact case, *Ann. Math.* **133** (1991), pp. 509–548.

[251] P. Vojta, Integral points on subvarieties of semiabelian varieties II, *Am. J. Math.* **121** (1999), pp. 283–313.

[252] M. Waldschmidt, Minorations de combinaisons linéaires de logarithmes de nombres algébriques, *Can. J. Math.* **45** (1993), pp. 176–224.

[253] M. Waldschmidt, *Diophantine Approximation on Linear Algebraic Groups: Transcendence Properties of the Exponential Function in Several Variables*, Berlin: Springer (2000).

[254] M. Watkins, Class numbers of imaginary quadratic fields, *Math. Comput.* **73** (2004), pp. 907–938.

[255] B. M. M. de Weger, Solving exponential Diophantine equations using lattice basis reduction algorithms, *J. Number Theory* **26** (1987), pp. 325–367.

[256] A. Wiles, Modular elliptic curves and Fermat's last theorem, *Ann. Math.* **141** (1995), pp. 443–551.

[257] J. Wolfart, Werte hypergeometrischer Funktionen, *Invent. Math.* **92** (1988), pp. 187–216.

[258] G. Wüstholz, Algebraische Unabhängigkeit von Werten von Funktionen, die gewissen Differentialgleichungen genügen, *J. Reine Angew. Math.* **317** (1980), pp. 102–119.

[259] G. Wüstholz, Über das Abelsche Analogon des Lindemannschen Satzes I, *Invent. Math.* **72** (1983), pp. 363–388.

[260] G. Wüstholz, Transzendenzeigenschaften von Perioden elliptischer Integrale, *J. Reine Angew. Math.* **354** (1984), pp. 164–174.

[261] G. Wüstholz, Zum Periodenproblem, *Invent. Math.* **78** (1984), pp. 381–391.

[262] G. Wüstholz, A new approach to Baker's theorem on linear forms in logarithms I, II. In: *Diophantine Approximation and Transcendence Theory*, ed. G. Wüstholz, Lecture Notes in Mathematics **1290**, Berlin: Springer (1987), pp. 189–202, 203–211. III. In: *New Advances in Transcendence Theory*, ed. A. Baker, Cambridge: Cambridge University Press (1988), pp. 399–410.

[263] G. Wüstholz, Multiplicity estimates on group varieties, *Ann. Math.* **129** (1989), pp. 471–500.

[264] G. Wüstholz, Algebraische Punkte auf analytischen Untergruppen algebraischer Gruppen, *Ann. Math.* **129** (1989), pp. 501–517.

[265] G. Wüstholz, Computations on commutative group varieties, *Arithmetic geometry, Sympos. Math.* **37** *(Cortona, 1994)*, Cambridge: Cambridge University Press (1997), pp. 279–300.

[266] Jing Yu, Analytic homomorphisms into Drinfeld modules, *Ann. Math.* **145** (1997), pp. 215–233.

[267] Kunrui Yu, Linear forms in the p-adic logarithms I, II, III, *Acta Arithmetica* **53** (1989), pp. 107–186; *Compos. Math.* **74** (1990), pp. 15–113; **76** (1990), p. 307; **91** (1994), pp. 241–276.

[268] Kunrui Yu, p-adic logarithmic forms and group varieties. I, II, *J. Reine Angew. Math.* **502** (1998), pp. 29–92; *Acta Arithmetica* **89** (1999), pp. 337–378.

[269] Kunrui Yu, Report on p-adic logarithmic forms. In: *A Panorama of Number Theory or the View from Baker's Garden*, ed. G. Wüstholz, Cambridge: Cambridge University Press (2002), pp. 11–25.

[270] O. Zariski, P. Samuel, *Commutative Algebra*, Vol. I, II, Berlin: Springer (1975).

Index